D4-02

Wildflowers of Door County

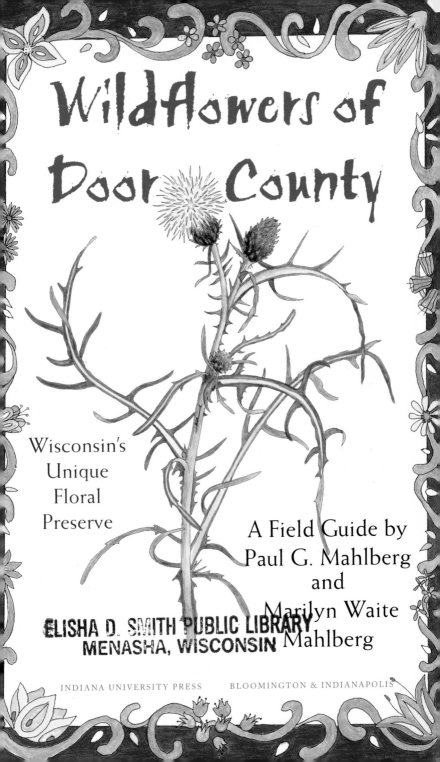

Wildflowers of Door County

Wisconsin's
Unique
Floral
Preserve

A Field Guide by
Paul G. Mahlberg
and
Marilyn Waite
Mahlberg

INDIANA UNIVERSITY PRESS   BLOOMINGTON & INDIANAPOLIS

This book is a publication of
Indiana University Press
601 North Morton Street
Bloomington, IN 47404-3797 USA
http://www.indiana.edu/~iupress
Telephone orders   800-842-6796
Fax orders   812-855-7931
Orders by e-mail   iuporder@indiana.edu

The paper used in this publication meets the minimum re-
quirements of American National Standard for Information Sci-
ences—Permanence of Paper for Printed Library Materials, ANSI
Z39.48-1984.
Printed in China

Library of Congress Cataloging-in-Publication Data

Mahlberg, Paul G.
Wildflowers of Door County : Wisconsin's unique floral preserve : a
field guide / by Paul G. Mahlberg and Marilyn Waite Mahlberg.
       p.   cm.
Includes bibliographical references (p.   )
ISBN 0-253-21453-X (pbk. : alk. paper)
1. Wildflowers—Wisconsin—Door County—Identification. 2. Wild
flowers—Wisconsin—Door County—Pictorial works. I. Mahlberg,
Marilyn Waite. II. Title.

QK194.M252001
582.13'09775'63—dc21   00-063251

1   2   3   4   5   06   05   04   03   02   01

To the kids
Melinda and Steven Fabel, Heidi and Michael Krall,

and the grandkids
Matthew and Jonathon Krall

Contents

Acknowledgments

THE AUTHORS ESPECIALLY wish to acknowledge an immense debt of gratitude to fellow botanist and friend Mary Standish, for her constant support and enthusiastic assistance in the preparation of this book. Mary helped in identifying and collecting plants for this book and for the herbarium at the University of Wisconsin in Madison.

We would like to express our thanks and appreciation to the Door County Parks Board and Parks Manager George Pinney for their encouragement and support.

We are deeply grateful to the many private land owners who allowed the authors access to their land. Because of them, many interesting and unusual collecting sites were found.

We wish to thank Prof. Hugh Iltis and Dr. Ted Cochrane of the Herbarium at the University of Wisconsin, Madison, for the assistance in verifying names of the specimens.

Special thanks to Bobbi Diehl and John Gallman, editors at Indiana University Press. John recognized the potential in producing a "Wisconsin" wildflower field guide and Bobbi took over from there and did all the editing. Our appreciation also to designer Sharon Sklar.

We are grateful to Patty Hooten and Fred Drescher of the Indiana University Biology Department for their valuable assistance in the preparation of this book

Lastly, our appreciation to good friends Kathryn and Joe Koehl and Simone Robbins for tolerating our constant obsession with the book.

Introduction

WISCONSIN'S DOOR COUNTY Peninsula is unique in possessing great ecological diversity, such as upland and boreal forest, bogs, swamps, sand and rock beaches, limestone escarpments and farmlands. Because of these numerous and varied ecosystems, a wide array of wildflowers occur in the county.

This book was written as a field guide to the wildflowers found in Door County throughout the growing season, from early spring through late fall. It is intended to help readers recognize and enjoy many of the wildflowers in their natural environment. Many of these plants have been recorded before in Door County, but we have observed some which appear to be new records, or which have not been detected for many decades. The range for most of these plants extends beyond the boundaries of our county to include the entire Great Lakes region, thus this guide can be used throughout the Midwest.

Our book is intended to stimulate your interest in the beauty of Door County and we hope that you will join us in the preservation of this most unusual area. With this in mind, we are contributing the profits from sales of our book to the Door County Land Trust, a nonprofit conservation organization dedicated to preserving important ecological areas for future generations to enjoy.

At present, there is no other floral, pictorial guide or book specific to Door County. Because we love Door County we decided to provide a permanent record of the abundant and beautiful wildflowers found here. We also wanted a guide that could be used and understood by the novice flower enthusiast. While nearly 400 local flowers are included in our book, you may come upon some not mentioned here. Indeed, it will be your challenge not only to find and enjoy those in our book, but to search further and try to find other species not included here.

All paintings were prepared from actual specimens of the flowering plant as observed in the field. Keep in mind that the color of flowers and the size of plants can vary depending on climate and the particular location of the plant. For many plants individual flowers were examined microscopically to obtain details of floral parts and accessory features such as hairs or bracts.

How to use our book. The most obvious aid to recognizing your quarry is flower color, and we have grouped plants accordingly. Under each color, plants are grouped in natural families, and the families arranged according to their level of complexity among Angiosperms, or flowering plants. The common and scientific names used in our book follow H. Gleason and A. Cronquist's "Manual of Vascular Plants of Northeastern United States and Adjacent Canada," 2nd ed., 1991. Each painting is accompanied with a brief description pointing out both general and distinguishing characters of the plant. The first sentence provides an overall description. It is followed with more details on plant organization, habitat, and period of flowering. The description is written in an abbreviated format to include as much information as possible about each plant. Compare several flowers, if available, when examining your specimen, because each may represent a different stage of maturity in the flowering process.

A glossary is included in the book defining various descriptive terms. Importantly, technical terminology is kept to a minimum. You will find it most desirable to own a small hand-held magnifying glass to examine various small flowers present on many plants.

We want you to enjoy the wildflowers. Do note that there are both federal and state ordinances to protect rare plants, and these are noted in our descriptions as: *Endangered,* meaning its continued existence is in jeopardy; *Threatened,* meaning it will likely become endangered in the near future; and *Special Concern,* meaning it is not abundant but further study is necessary before classifying it in either above category. So do admire them, photograph them, and let them remain intact for others to enjoy. Door County is resplendent with wildflowers; help us keep it this way.

LEAF BLADE

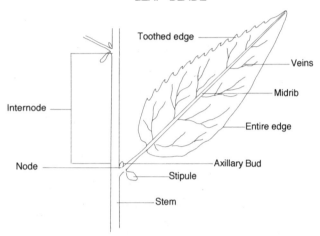

Toothed edge

Veins

Midrib

Internode

Entire edge

Axillary Bud

Node

Stipule

Stem

LEAF ARRANGEMENT

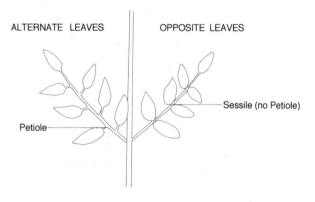

ALTERNATE LEAVES

OPPOSITE LEAVES

Sessile (no Petiole)

Petiole

LEAF SHAPES

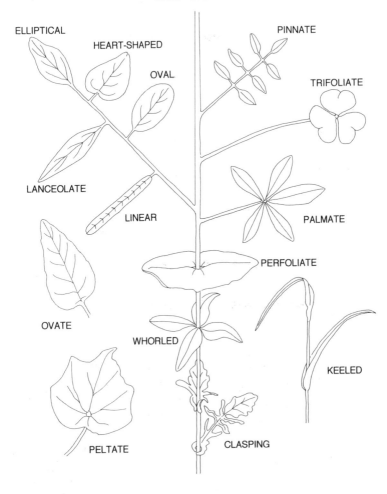

ELLIPTICAL

HEART-SHAPED

OVAL

PINNATE

TRIFOLIATE

LANCEOLATE

LINEAR

PALMATE

OVATE

PERFOLIATE

WHORLED

KEELED

PELTATE

CLASPING

I L L U S T R A T I O N S

FLOWER TYPES

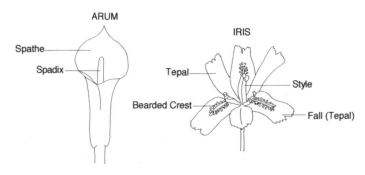

IRREGULAR FLOWERS

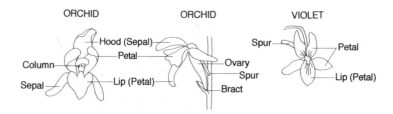

REGULAR FLOWERS

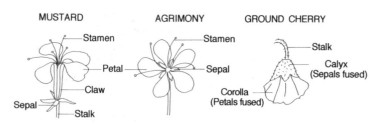

FLOWER PARTS

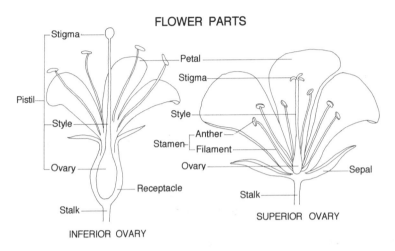

INFERIOR OVARY

SUPERIOR OVARY

FLOWER TYPES

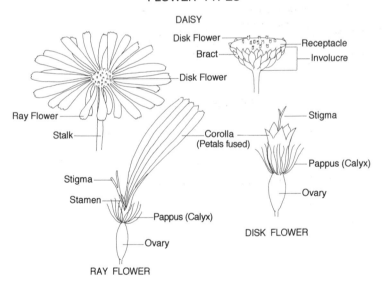

DAISY

RAY FLOWER

DISK FLOWER

Glossary

ACHENE. A small, dry, usually one-seeded fruit.

ACRID. Bitter, stinging, irritating to skin, or body if ingested.

ACUMINATE. Tapering to a narrow point.

AGGREGATE FRUIT. Fruit derived from 2 or more pistils of one flower, as a raspberry.

ALTERNATE. Positioned singly at each node, such as leaves on a stem.

AMPHIBIOUS. Lives or grows both in water and on land, such as those plants growing on the land after water has receded.

ANNUAL. A plant that lives for and completes its life cycle in one season of growth.

ANTHER. That part of stamen bearing pollen.

APPRESSED. Lying against or close to a structure, as hairs growing close to the stem or leaf.

ARIL. A specialized usually fleshy outgrowth on a seed; often attractive to other organisms as food to aid seed dispersal.

AWN. A narrow tip such as a bristle or very sharp tip on a leaf.

AXIL. The position or angle formed between a leaf and the stem.

AXILLARY BUD. A bud formed in the axil.

BARBED. Possessing short, firm points.

BEAK. A prominent tip or appendage on a structure, such as a seed.

BEARDED. Possessing a cluster or tuft of long hairs.

BERRY. A fleshy fruit from a single pistil usually with several or more seeds.

BIENNIAL. A plant requiring two growing seasons to complete

its life cycle, usually producing a basal rosette the first season and the flowering axis the second year of growth.

BLADE. Flat expanded portion of a leaf, petal or sepal.

BRACT. Small modified leaf usually growing at the base of a leaf, flower or flower cluster.

BRISTLE. A stiff hair.

BUD. An unexpanded leafy shoot or flower.

BULB. An underground bud with scaly fleshy leaves

BULBOUS. Swollen part on an organ; like a bulb.

CALCAREOUS. A limey soil, rich in calcium, especially calcium carbonates.

CALYX. The outer ring of flower parts, collectively the sepals.

CAMPANULATE. Bell-shaped, to describe a corolla or calyx.

CAPILLARY. Shaped like a hair.

CAPITATE. The form of a head.

CAPSULE. A dry, dehiscent fruit consisting of more than one carpel or section.

CATKIN. A short compact spike of flowers, each flower subtended by a bract; cone-like in appearance.

CHAFF. Dry, thin scales, such as the bracts on the receptacle of some daisy flower-heads; modified leaves.

CILIATE. Fringed with hair.

CLASPING. Base of a leaf that partially surrounds the stem.

CLAW. The narrow, lower portion of sepals or petals below their expanded blade.

CLEFT. Deeply cut or lobed.

COLONIAL. Forming colonies by many offshoot plants developing from one plant.

COLUMN. The united filaments of stamens, or of filaments and style as in orchids.

COMPLETE FLOWER. Flower with all parts including sepals, petals, stamens and ovary.

COMPOUND LEAF. A leaf with two or more separate leaflets.

CONICAL. The form of a cone.

CORM. A thickened usually vertical underground stem without fleshy leaves.

COROLLA. The inner ring of flower parts, collectively the petals.

CORONA. A ring present at the throat of a flower formed by structural modification of a portion of the petals or stamens.

CRESTED. Bearing evident ridges or raised thickened areas on its surface.

CYATHIUM. A very specialized (false) flower of a *Euphorbia* consisting of a cup-shaped involucre, often with petal-like appendages. Modified flowers occur in the involucre: a single central female flower consists of a 3-parted ovary on a short stalk, and surrounding male flowers each consist of 1 stamen. After fertilization the stalk elongates and lifts the ovary out of the cyathium.

DISCOID. The form of a disk, as in Daisy family when all flowers are the disk type.

DISK. The central portion of a flower-head in the Daisy family.

DISK FLOWER. The specialized central tubular flowers in the disk of a flower-head.

DISSECTED. Much divided into very small or fine segments.

DOWN. Covered with fine hairs.

DRUPE. A fleshy fruit with a central firm stony part enclosing a single seed, and a firm skin surrounding the fleshy region, as a peach.

DRUPELET. Diminutive form of drupe.

ELLIPTICAL. Broadest near the middle and generally tapering toward both ends.

ENTIRE. Having no teeth, a continuous unbroken margin.

EYE. A prominent structure or mark at center of a flower, as at its throat.

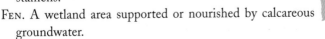

FEMALE FLOWER. A flower bearing one or more ovaries, but no stamens.

FEN. A wetland area supported or nourished by calcareous groundwater.

FILAMENT. Stalk supporting the anther of a stamen.

FL. Flowering period.

FLOWER-HEAD. The daisy type flower with disk and/or ray flowers.

FOLLICLE. A dry fruit with many seeds formed from a pistil and opening along a suture.

FRINGE. Furnished with hairs.

FRUIT. A ripened ovary and any other surrounding structures.

GLAND. A structure or protuberance that is specialized to produce sticky substances.

GLOBOSE. Spherical in shape.

GYNOSTEGIUM. Specialized flower organization in Asclepiadaceae in which the united filaments surround the style and the anthers are attached to a large disk-like fleshy stigma.

HABITAT. The area or place in an ecosystem where a plant lives.

HIP. Cup-shaped fruit of the rose.

HOOD. Cup-like structure formed from anther filament in Asclepiadaceae, and vary in structure, shape and size.

HORN. Curved, pointed structure formed from hood in Asclepiadaceae, and varies in structure, shape and size.

HYPANTHIUM. Specialized swollen ring at base of the ovary in some flowers at which petals and sepals are attached.

INFERIOR OVARY. The sepals, petals and stamens are positioned at the top of the ovary.

INFLATED. Puffed out or bladdery.

INFLORESCENCE. The cluster and arrangement of flowers on the stem.

INTERNODE. The stem region between two nodes.

INTRODUCED. A plant not native to the region covered in this guide.

INVOLUCRAL. Term applied to components of the involucre.

INVOLUCRE. The set of bracts and the region beneath the ray and/or disk flowers in the Asteraceae, in particular; a structure surrounding the base of another structure.

IRREGULAR. A non-symmetrical flower, one in which petals or sepals are dissimilar in shape.

KEEL. A ridge along a structure, as the longitudinal ridge along a partly folded leaf.

LANCEOLATE. Longer than broad, and widest below the middle and tapering at one or both ends.

LEAFLET. One of the separate or divided parts of a leaf.

LINEAR. Long, narrow, with parallel sides, as the shape of a leaf.

LIP. One part of an unequally divided corolla or calyx differing in size, shape or other feature.

LOBE. One or more of the projecting segments, often rounded, of a leaf or flower.

MALE FLOWER. A flower bearing stamens, but no functional ovary.

MARSH. A wet, treeless area.

MEADOW. A treeless area less wet than a marsh.

MIDRIB. Main vein of a leaf.

MULTIPLE FRUIT. Fruit formed from several to many fused flowers, as a pineapple.

MYCOTROPHIC. A physiological interrelationship between a plant root and a fungus in the soil.

NECTAREOUS. Bearing nectar.

NECTARY. A specialized structure usually at the base of a flower part that produces nectar.

NERVE. Refers to veins of a leaf, bract or other structure.

NODE. The place on a stem where the leaf is attached; contrast with the internode.

NUT. A dry, indehiscent hard fruit; a thick-walled achene.

NUTLET. Small nut; used to indicate one part of a nut fruit composed of several joined together.

OBLONG. Uniformly longer than wide, as a rectangle.

OPPOSITE. Positioned directly across from each other, as leaflets on a leaf petiole.

OVAL. Egg-shaped; here as for a leaf, the widest end could be either toward the top or base.

OVARY. The structure which encloses one or more ovules; the basal part of a pistil with ovules.

OVATE. Egg-shaped with broad end toward base.

PALATE. The raised often thickened area of a calyx or corolla part.

PALMATE. With leaflets or lobes radiating from a common point, as on a hand.

PAPPUS. The modified calyx on an (inferior) ovary, as for Asteraceae, consisting of hairs, bristles or scales.

PARASITE. A green or non-green plant that obtains its nutrients and water from another living plant or organism.

PELTATE. A leaf with the petiole attached to the underside of the blade, umbrella-like.

PERENNIAL. A plant that usually produces flowers every year, and lives for several to many years.

PERFOLIATE. A leaf for which the basal margins surround the stem, so that the stem appears to pass through the leaf.

PERIANTH. The calyx and corolla collectively; often used when both are very similar.

PERSISTENT. Remaining attached after normal function is completed.

PETAL. One of the colored or white parts of a flower; one of the parts of the corolla.

PETIOLE. A leaf stalk.

PINNATE. With leaflets or lobes formed on both sides along an axis, as along a petiole.

PISTIL. The female or seed-bearing structure of a flower including the ovary, style and stigma.

PLUMOSE. Feathery, as with dense hairs, or pinnately arranged bristles.

POD. A dry, dehiscent fruit, such as pea, which opens by splitting longitudinally.

POLLEN. The (pollen) grains produced in the anther that contain the male gametes.

POLLINATION. The transfer of pollen from the anther to the stigma.

POME. A fleshy fruit with a soft core, such as an apple, usually derived from an inferior ovary.

RAY. The margin-flowers on a flower-head, in the Daisy family, with a long, flat modified corolla; also called ligulate flower.

RECEPTACLE. The widened basal portion of a stem related to a flower or flower-head; in the Daisy family the ray and disk flowers are attached to a receptacle.

REFLEXED. Bent backwards.

REGULAR. A flower in which its parts, particularly the sepals and petals, are similar in size and shape.

RETICULATE. A network, such as the veins of a leaf.

RHIZOME. An underground, horizontally growing stem.

ROSETTE. A dense cluster of leaves arranged in a circle at the base of a plant.

RUNNER. See stolon.

SAPROPHYTE. A non-green plant that lives on decaying organic matter; the plant is not parasitic, nor does it manufacture its own food.

SCALE LEAF. Modified leaf.

SEED. The mature ovule, including the embryo and surrounding parts.

SEPAL. One of the outer parts of a flower, usually green; one of the parts of the calyx.

SESSILE. Attached without a petiole.

SHEATH. An organ, such as a leaf or stipule, that partly or entirely surrounds another organ, as in the Smartweed family.

SHOOT. A new stem with its leaves, such as that forming in a leaf axil.

SMOOTH. Not hairy.

SPADIX. Thick stem region bearing many crowded stalkless flowers, as in Arum family.

SPATHE. Large white or colored bract surrounding, or partially surrounding, a flower cluster, as in Arum family.

SPATULATE. Shaped similar to a spatula.

SPIKE. An elongated inflorescence with sessile or nearly sessile flowers.

SPUR. A hollow tubular extension of a petal, usually, which collects nectar.

STALK. The stem of a flower.

STAMEN. Pollen-bearing organ of the flower consisting of anther and filament. Broadly, the male part of flower.

STAMINODE. A non-functional stamen.

STEM. Main axis of a plant.

STIGMA. The pollen-receiving portion of the pistil.

STIPULE. Modified leaf, usually pair, at the base of a leaf of many plants.

STOLON. A prostrate stem growing on the surface of the ground and rooting at its nodes.

STRIATE. Marked with parallel ridges or lines.

STYLE. The stalk which connects the stigma to the ovary.

SUBTEND. To occur immediately below a structure, such as a bract below a flower.

SUPERIOR OVARY. The sepals, petals and stamens are positioned at the base of the ovary.

SWAMP. A wet, wooded area.

TENDRIL. A thin, twining tactile organ, such as a modified leaf, specialized to coil and attach to a host for support.

TEPAL. Undifferentiated, or similar in character, as when sepals and petal look similar.

THROAT. The opening of a corolla or calyx, that region between the tube and the lips.

TOOTHED. To have many often pointed indentations along the
margin, as on a leaf.

TRIFOLIATE. Three leaflets, as in clover.

UMBEL. A cluster of flowers with their equal stalks radiating
from a common point.

UMBELLET. A small umbel on a larger one.

UTRICLE. An inflated, but small, thin-walled one-seed fruit.

VEIN. A vascular bundle as evident in a leaf or petal.

WHORL. With three or more structures, as leaves, radiating
from a node.

WING. A flat extension from the side of a structure, as on a
petiole.

WOOLLY. Finely hairy.

Yellow

FLOWERS

YELLOW WATER-LILY
Nuphar variegata

WATER-LILY FAMILY   NYMPHAEACEAE

Perennial aquatic, 4' tall, bearing
large floating yellow flower at top of
long leafless stalk. Flower 3" wide,
sepals 5–6, showy, yellow or green
with maroon inner basal region.
Petals many, small, fleshy at base
of stamens. Many stamens in
several rings encircle several
united pistils. Ovary superior, 1–
many pistils, large discoid stigma.
Fruit leathery berry. Floating leaf
blade 15" wide, heart-shaped base,
basal lobes often overlap, smooth, entire,
from underwater rhizome on long petiole.
Habitat: ponds, quiet water. FL: May–
September.

MARSH-MARIGOLD
Caltha palustris

BUTTERCUP FAMILY   RANUNCULACEAE

Perennial, 2' tall, bearing large solitary shiny
yellow flower on long stalk near top of branch-
ing leafy stem. Flower 1¹/₂" wide, petals
none. Sepals 5–9, yellow, oval.
Stamens many, short. Each
ovary superior, several pistils,
short style. Fruit follicles in
whorl, many seeds. Stems smooth,
hollow, branched above. Basal
leaves 7" wide, toothed, heart-
shaped base, long-petioled;
upper leaves smaller, linear
base, shorter petioles, toothed;
alternate. Poisonous if eaten.
Habitat: swamps, marshes, wet
meadows. FL: April–June.

SMALL-FLOWERED CROWFOOT
Ranunculus abortivus
BUTTERCUP FAMILY   RANUNCULACEAE

Biennial, 2' tall, bearing terminal inconspicuous yellow flowers with drooping sepals on branched leafy finely hairy stems. Flower $1/4$" wide, petals 5, narrow, spreading (right). Sepals 5, oval, pointed, reflexed, longer than petals. Stamens many. Ovary superior, many pistils, short styles. Fruit globose cluster of beaked achenes (left). Basal leaves $1^1/_2$" wide, kidney-shaped, scalloped margins, long-petioled. Stem leaves 3–5 lobed, toothed, petioled except uppermost; alternate. Poisonous if eaten. Habitat: moist woods, streambanks. FL: April–June.

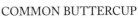

COMMON BUTTERCUP
Ranunculus acris
BUTTERCUP FAMILY   RANUNCULACEAE

Perennial, 3' tall, bearing glossy yellow flower at top of branched leafy hairy stem. Flower 1" wide, petals 5, heart-shaped, spreading, notched tips. Sepals 5, short. Stamens many, yellow. Ovaries superior, many pistils. Fruit globose cluster of flat achenes with curved beak, winged edges. Hairs appressed against stem, or sometimes spreading out from stem. Basal leaves 4" wide, long-petioled, hairy, blade deeply 3–7 palmately lobed, these further dissected. Upper leaves smaller, 3-parted, scattered along stem, terminal lobe sessile. Contains acrid juice. Habitat: fields, meadows; introduced. FL: May–September.

LESSER CELANDINE
Ranunculus ficaria
BUTTERCUP FAMILY   RANUNCULACEAE

Perennial, 1' tall, bearing single large glossy yellow
flower at top of stalk on leafy stem. Flower $1^1/_2$"
wide, petals 8–12 glossy, elongated $^5/_8$" long, each
with basal nectary. Sepals 3, elongated, green,
short. Stamens many, yellow. Ovaries superior,
many pistils, short style. Fruit globose head of
plump beakless achenes. Stem smooth with
flowers on short shoots. Basal leaves $1^1/_2$"
wide, small lobes, heart-shaped base, long-
petioled, blunt-tipped. Leaves on stalk
smaller, opposite or alternate. Habitat:
wet meadows, roadsides; introduced. FL: May–July.

YELLOW WATER-CROWFOOT
Ranunculus flabellaris
BUTTERCUP FAMILY   RANUNCULACEAE

Perennial aquatic, 2' long, bearing 1-few yellow
flowers at top of emerged stalk from submerged stem with feathery
leaves. Flower 1" wide, petals
5, spreading, basal nectaries.
Sepals 5, half length of petals.
Stamens many, yellow. Several
superior ovaries, several pistils. Fruit
globose head of plump achenes, curved
beak, corky margin. Flower stalk long,
each subtended by bract. Stem usually
submerged, hollow, branched. Leaves
$1^1/_2$" long, usually submerged, much
dissected into fine filaments,
petiole shorter than stipule.
Habitat: quiet pools, muddy shores;
rare. FL: April–June.

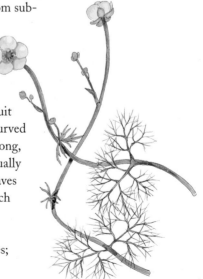

CREEPING SPEARWORT
Ranunculus flammula
BUTTERCUP FAMILY   RANUNCULACEAE

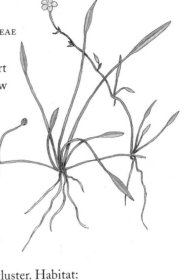

Creeping perennial, 4" tall, bearing yellow flower at top of stalk on short leafy stem with linear or very narrow spatulate leaves. Flower $^1/_4$" wide, petals 5, oval, spreading, basal nectary. Sepals 5, half length of petals. Stamens, many, yellow. Ovary superior, several pistils. Fruit globose head of winged achenes, erect beak. Stem runners root at nodes to produce new leaf and flower axes. Leaves 4" long × $^1/_4$" or less wide, spatulate, long petiole, entire, smooth; basal cluster. Habitat: wet rocky sandy shoreline, not submerged; rare. FL: June–August.

HISPID BUTTERCUP
Ranunculus hispidus var. *caricetorium*
BUTTERCUP FAMILY   RANUNCULACEAE

Perennial, 18" tall, bearing glossy yellow flowers at top of branched leafy hairy stems. Flower $1^1/_2$" wide, petals 5 oval, spreading, tip notched, green nectary at base of petal. Sepals 5, small. Stamens many, yellow. Ovaries superior, many pistils. Fruit globose cluster of flat achenes, straight beak, ridged margin. Basal leaves 3-lobed, petioled, hairy. Upper leaves palmately lobed, further divided with cuts and teeth, or strap-like. Poisonous if eaten. Habitat: wet woods, creeksides. FL: April–June.
Variety. *nitidus* similar but reclining, hollow stem, leaflets short-petioled, seed margin winged.

HOOKED CROWFOOT
Ranunculus recurvatus

BUTTERCUP FAMILY   RANUNCULACEAE

Perennial, 18" tall, bearing inconspicuous yellow flowers with drooping sepals at tips of leafy hairy stems. Flower $^1/_4$" wide, petals 5, spreading, narrow. Sepals 5 or more, reflexed, short and long, hairy. Stamens many, yellow. Ovaries superior, many pistils. Fruit globose cluster of plump achenes with hooked beak. Basal leaves $2^1/_4$" wide, 3-lobed, toothed, petioled. Stem leaves 2" wide, toothed, 3-cleft with lateral leaflets lobed, pointed. Poisonous if eaten. Habitat: dry or moist woods. FL: May–June.

CURSED CROWFOOT
Ranunculus sceleratus

BUTTERCUP FAMILY   RANUNCULACEAE

Perennial, 2' tall, bearing small yellow flowers with drooping sepals at top of branched leafy hollow stems. Flower $^3/_{16}$" wide, petals 5 spreading, round. Sepals 5 reflexed, short. Stamens many, yellow anthers. Ovaries superior, many pistils. Fruit cylindrical cluster, plump achenes, beakless. Flower subtended by leaf. Stem smooth. Basal leaves petioled, deeply 3-cleft, segments again deeply lobed. Upper leaves small, linear-elongated, smooth, sessile. Juice is acrid. Habitat: wet ditches, marshes. FL: May–August.

GOLDEN CORYDALIS
Corydalis aurea
FUMITORY FAMILY   FUMARIACEAE

Biennial, 4' tall, bearing several
fragrant irregular yellow
flowers in axils of dis-
sected leaves on smooth
stem. Flower $^5/_8$" long,
petals 4. Upper crested
petal tip curled upward,
slightly toothed; lower
petal tongue-like, keeled;
two narrow lateral petals fused at tip
enclosing anthers and stigma. Stamens 6.
Ovary superior, 1 style, T-shaped stigma. Fruit
elongated drooping capsule, black ariled seeds.
Flower subtended by persistent bract. Leaves
6" long, petioled, 2×-pinnate, each
leaflet lobed, entire, smooth; alternate.
Habitat: rocky wooded banks,
clearings. FL: May–July.

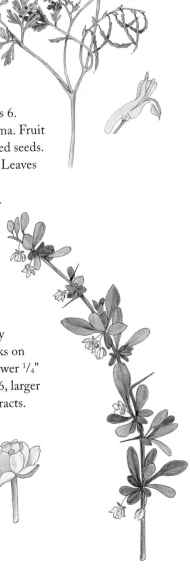

JAPANESE BARBERRY
Berberis thunbergii
BARBERRY FAMILY   BERBERIDACEAE

Perennial, 4' tall, bearing small axillary
clusters of yellow flowers on long stalks on
leafy stem with long sharp spines. Flower $^1/_4$"
wide, petals 6, oval, glandular. Sepals 6, larger
than petals, subtended by 2–3 small bracts.
Stamens 6. Ovary superior, short
stigma. Fruit 1-seeded elliptic
red berry. Stem shrubby, leaf
cluster on spur branches, $^3/_8$"
pink spines. Leaves 1" long,
oval-oblong, entire, smooth,
narrowed toward base, petioled;
alternate. Habitat: rocky woods,
roadsides; introduced. FL: May.

KALM'S ST. JOHN'S-WORT
Hypericum kalmianum

MANGOSTEEN FAMILY   CLUSIACEAE

Perennial, 3' tall, bearing several large yellow flowers at top of branched leafy stem. Flower 1" wide, petals 5, spreading; become upright in older flowers. Sepals 5, often folded along middle, persistent. Stamens many, yellow. Ovary superior, 5-celled, 5 united styles. Fruit oval capsule, sharp stylar beak. Stem shrubby, rough. Leaves $1^1/_2$" long, linear-oblong, thick, scalloped margins rolled under, sessile; appear clustered at nodes, but opposite. Habitat: calcareous soil, lakeshores. FL: July–August.

COMMON ST. JOHN'S-WORT
Hypericum perforatum

MANGOSTEEN FAMILY   CLUSIACEAE

Perennial, 3' tall, bearing large yellow flowers at tips of leafy stems on branched plant. Flower 1" wide, petals 5, spreading, with red-black dots along margin. Sepals 5, short, pointed. Stamens many, clustered around pistil. Ovary superior, 1 pistil, 3 long styles, globose stigma. Fruit capsule with persistent sepals, many seeds. Stem angled, ridged. Leaves $1^1/_2$" long, narrow-oblong, entire, sessile; opposite. Many oil glands evident as translucent dots on leaf when held to light. Habitat: sunny fields, meadows; introduced. FL: June–August.

DOWNY YELLOW FOREST-VIOLET
Viola pubescens
VIOLET FAMILY   VIOLACEAE

Perennial, 16" tall, bearing
nodding irregular yellow
flowers on long stalk in axil
near top of leafy finely hairy
stem. Flower ³/₄" wide, petals
5, spreading, purple veins near
base, lateral 2 bearded. Sepals 5,
small, lanceolate. Stamens 5,
extend to throat. Ovary
superior, 1 bearded style. Fruit
elongated capsule, brown seeds.
Stalk with small bracts, same height as leaves.
Petioled leaves 3" long, heart-shaped blade,
toothed, pointed; long pointed stipule.
Habitat: rich dry woods. FL: May–June.
Smooth Y. F., *V. pensylanica,* similar but not hairy.
Habitat: rich moist woods.

ROUND-LEAVED YELLOW VIOLET
Viola rotundifolia
VIOLET FAMILY   VIOLACEAE

Perennial, 5" tall, bearing irregular nodding yellow
flowers on stalks separate from leaves. Flower ¹/₂"
wide, petals 5, upper 2 reflexed, lateral
2 bearded with yellow hairs and 1
dark vein, burgundy veins mark
lower petal. Sepals 5, short, pointed.
Stamens 5, extend to throat. Ovary
superior, 1 pistil, 1 style. Fruit ¹/₄"
long purple-dotted capsule,
cream-colored seeds. Leaves 2"
wide, heart-shaped, toothed, long
finely hairy petiole arises from
stout rhizome. Habitat: rich
conifer woods. FL: April–
May.

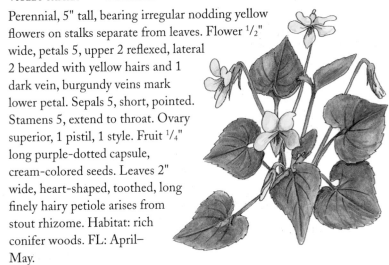

YELLOW ROCKET
Barbarea vulgaris

Mustard family   Brassicaceae

Biennial, 2' tall, bearing dense terminal clusters
of yellow flowers at top and in axils of leafy
stem. Flower $^1/_3$" wide, petals 4, spatulate, taper
to claw, form cross above 4 pointed sepals.
Stamens 6, two short with basal glands, 4
long separated by erect glands. Ovary
superior, short style. Fruit $1^1/_2$" long,
linear 4-angled pod, long stalk, short
beak, black seeds. Leaves petioled, entire,
smooth, pinnate, 1–4 lobes with large terminal
lobe. Lower leaves petioled, upper clasp stem;
alternate. Habitat: cultivated fields, disturbed
areas; introduced. FL: April–August.

BLACK MUSTARD
Brassica nigra

Mustard family   Brassicaceae

Annual, 3' tall, bearing terminal clusters of yellow flowers at top of
leafy hairy stem with large-lobed basal leaves. Flower $^1/_3$" wide, petals
4, spatulate, taper to claw. Sepals 4,
narrow, reflexed. Stamens 6,
two short, 4 long; anthers
yellow. Ovary superior,
globose stigma. Fruit $^3/_4$"
long pod, erect, 1 main vein,
persistent beak; transversely
jointed near beak base. Seeds
brown, pitted. Pod
grows before petals
fall. Basal leaves
with large terminal
lobe and 4 lateral
lobes; upper leaves lan-
ceolate, toothed, not lobed.
Habitat: pastures, roadsides;
introduced. FL: May–July.

HERB-SOPHIA
Descurainia sophia
<small>MUSTARD FAMILY   BRASSICACEAE</small>

Annual, 30" tall, bearing terminal cluster of small yellow flowers at top of leafy stem with fern-like leaves. Flower $1/8$" wide, petals 4, elongate, small. Sepals 4, longer than petals, oblong, edges yellow. Stamens 6, about equal length, yellow anthers. Ovary superior, 1 pistil, yellow-green style, globose stigma. Fruit long-stalked 4-angled pod, erect, persistent style, seeds in 1 row. Stem branched, hairs branched. Leaves 1–3× pinnately divided into narrow leaflets. Habitat: disturbed areas, fields; introduced. FL: May–July.

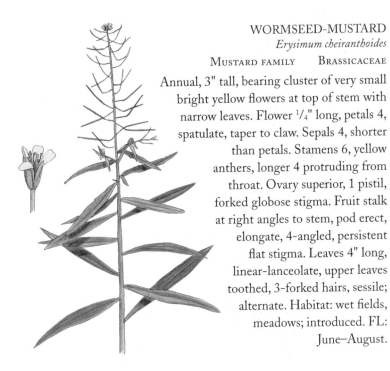

WORMSEED-MUSTARD
Erysimum cheiranthoides
<small>MUSTARD FAMILY   BRASSICACEAE</small>

Annual, 3" tall, bearing cluster of very small bright yellow flowers at top of stem with narrow leaves. Flower $1/4$" long, petals 4, spatulate, taper to claw. Sepals 4, shorter than petals. Stamens 6, yellow anthers, longer 4 protruding from throat. Ovary superior, 1 pistil, forked globose stigma. Fruit stalk at right angles to stem, pod erect, elongate, 4-angled, persistent flat stigma. Leaves 4" long, linear-lanceolate, upper leaves toothed, 3-forked hairs, sessile; alternate. Habitat: wet fields, meadows; introduced. FL: June–August.

COMMON YELLOW-CRESS
Rorippa palustris

MUSTARD FAMILY   BRASSICACEAE

Biennial, 3' tall, bearing long clusters of small yellow flowers at top and in axils of leafy hairy stems. Flower $3/16$" wide, petals 4, oval. Sepals 4, yellowish, equal to or longer than petals. Stamens 6, two short, 4 protrude above petals, yellow anthers. Ovary superior, 1 pistil, fuzzy capitate stigma. Fruit plump pod $3/8$" long, attached at right angles to stem, many brown pitted seeds. Leaves 6" long, pinnately lobed, marginal teeth, smooth or hairy, sessile, lobed base clasps stem, midrib a white line; alternate. Habitat: ponds, lakes; introduced. FL: May–October.

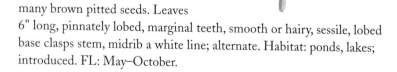

TUMBLING MUSTARD
Sisymbrium altissimum

MUSTARD FAMILY   BRASSICACEAE

Biennial, 3' tall, bearing long cluster of small pale yellow flowers at top of stem with pinnately lobed leaves. Flower $3/4$" wide, petals 4, spatulate, clawed. Sepals 4, short. Stamens 6, two short with basal glands. Ovary superior, 1 pistil, short style. Fruit 4" long, slender, 3-nerved pod, stalk as thick as pod, elongates with petals at base, pods spread outward from stem. First year rosette leaves 5–8 pinnate lobes. Stem leaves 4" long, 2–7 pinnate lobes, lobes round-toothed, smooth, petioled; alternate. Habitat: disturbed areas; introduced. FL: May–August.

HEDGE-MUSTARD
Sisymbrium officinale
MUSTARD FAMILY   BRASSICACEAE

Annual, 3' tall, bearing long cluster of small bright yellow flowers at top of stiffly branched stem with dagger-like leaves. Flower $1/8$" wide, petals 4, spatulate, notched tip, taper to claw. Sepals 4, half length of petals, red-brown. Stamens 6, yellow anthers, two short with basal glands. Ovary superior, 1 cylindrical pistil, short style. Fruit $5/8$" long, erect, adjacent to stem, angular, 1 row of smooth seeds. Stem mottled red, upper half branched, finely hairy. Leaves 6" long, dagger-like pinnately lobed, toothed or wavy, lower petioled and upper sessile; alternate. Habitat: disturbed areas; introduced. FL: May–October.

WHORLED LOOSESTRIFE
Lysimachia quadrifolia
PRIMROSE FAMILY   PRIMULACEAE

Perennial, 30" tall, bearing several long-stalked yellow star-shaped flowers in axils of whorled leaves near stem tip. Flower $1/2$" wide, on long slender stalk, corolla 5-lobed, tips pointed, throat of spreading lobes marked with red. Calyx lobes 5, short, spreading. Stamens 5, glandular filament, form tube around ovary base. Ovary superior, 1 pistil, 1 long style, stigma above stamens. Fruit capsule, persistent beak. Leaves 4" long, narrow or broadly lanceolate, smooth, entire, sessile, whorls of 3–6 on smooth stem. Habitat: moist or dry open woods. FL: June–August.

SWAMP-LOOSESTRIFE
Lysimachia thrysiflora

PRIMROSE FAMILY   PRIMULACEAE

Perennial, 30" tall, bearing
clusters of yellow flowers in axils
of middle leaves along leafy stem.
Flower $\frac{1}{4}$" long, corolla tube short,
corolla lobes long, narrow, with dark
dots. Calyx 6 short lobes. Stamens 6,
longer than petals, yellow anthers on
slender filaments. Ovary superior, 1 pistil,
1 long style. Fruit globose
capsule. Stem smooth,
somewhat woody. Leaves $4\frac{1}{2}$" long,
narrow, lanceolate, smooth, marked with dots,
pointed, sessile; opposite. Habitat: wet woods,
swamps. FL: May–July.

EASTERN BLACK CURRANT
Ribes americanum

GOOSEBERRY FAMILY   GROSSULARIACEAE

Perennial, 4' tall, bearing more
than 3 yellow flowers on flower
stalks in axils of leafy thornless
stem with 3-lobed leaves.
Flower smooth cup $\frac{1}{4}$" wide, 5
round petals. Sepals 5, smooth,
longer than petals, appear part of
cup. Stamens 5, short. Ovary
inferior, short lobed style. Fruit
smooth black berry. Stalk equal
to or shorter than subtending
bract. Leaves 3" long, 3–5
palmately lobed, toothed,
glands on underside, hairs at
petiole base; alternate. Habitat: moist
woods. FL: April–May.

Western B. C., *R. hudsonianum*, flower stalk
longer than bract, sepals hairy. *Special Concern.* Habitat: swamps.

BUFFALO-CURRANT
Ribes odoratum

GOOSEBERRY FAMILY   GROSSULARIACEAE

Perennial, 4' tall, bearing 3 or more clove-scented yellow flowers on stalk in axils on leafy thornless stem. Flower ³/₈" wide, petals 5, very short. Sepals 5, yellow, petal-like, appear as part of smooth long "calyx" tube above ovary. Stamens 5, short. Ovary inferior, short style. Fruit yellow or black. Leaf blade oval, tip with 3 lobes, no glands on underside, finely hairy margin; alternate. Habitat: rocky hillsides; introduced. FL: April–July. Swamp Red C., *R. triste,* similar but flowers non-scented, stalk glandular, fruit red. Habitat: bogs.

GOLDEN CARPET
Sedum acre

STONECROP FAMILY   CRASSULACEAE

Perennial, 4" tall, bearing small cluster of several star-shaped yellow flowers at top of succulent leafy stem. Flower ¹/₂" wide, petals 5, spreading, pointed, united at very base. Sepals 5, small. Stamens as many as 10, long filaments, yellow. Ovary superior, 5 star-like short stigma-styles. Fruit capsule, small brown seeds. Stem creeping, matted, with short erect leafy flower stems. Leaves ¹/₄" long, succulent, overlap each other, old leaves persist; alternate. Habitat: open rocky sandy area; introduced. FL: June–July.

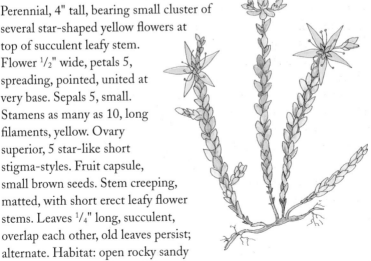

COMMON AGRIMONY
Agrimonia gryposepala

ROSE FAMILY   ROSACEAE

Perennial, 3' tall,
bearing wand-like
glandular spike of
small yellow flowers at top of
leafy hairy stem with pinnate
leaves. Flower $1/4$" wide, petals 5,
spreading. Sepals 5, small.
Stamens 5, long. Ovary inferior,
2 pistils, styles not jointed near
middle. Fruit achene, oval,
hooked hairs spreading
outwardly that attach to fur,
clothing. Leaves pinnate with 5–7
larger leaflets 3" long interspersed
with small leaflets along long petiole,
coarsely toothed, many-veined, stipules;
alternate. Crushed stem fragrant. Habitat:
rich woods, thickets. FL: July–August.

YELLOW AVENS
Geum aleppicum

ROSE FAMILY   ROSACEAE

Perennial, 2' tall, bearing small cluster of
nodding bell-shaped yellow flowers
at top of leafy hairy stem with
pinnate leaves. Flower $3/4$" wide,
petals 5, broad, spreading. Sepals 5,
narrow, pointed, become reflexed. Stamens
10, yellow anthers. Ovaries many on
receptacle, long jointed style. Fruit hairy
achenes; bristles among achenes on
receptacle. Basal leaves pinnate, upper
3 or 5 leaflets similar, 3" long, smaller
below, toothed, petioled. Stem leaves
smaller, 3-foliate; alternate. Habitat:
swamps, wet meadows. FL: May–July.

Perennial, 1' tall, bearing one yellow
flower at top of long stalk from leafy
stolon. Flower $3/4$" wide, petals 5, spatulate, spreading. Sepals 5. Bracts 5, narrow,
alternate with sepals. Stamens short,
black. Ovaries superior, many on
central dome. Fruit furrowed
achenes. Stolon forms leaf or flower
axes at nodes. Leaves 6" long,
pinnate, many leaflets, leaflet size
increases upward, toothed, silver
hairy underside, stipules. Habitat:
wet beaches. FL: May–October.
Old-field Five-fingers, *P. simplex,* similar but shorter, smaller flowers,
5-palmate oval leaflets. Habitat: dry open areas.

SHRUBBY FIVE-FINGERS
Potentilla fruticosa
Rose family   Rosaceae

Perennial, 3' tall, bearing 1 or few
large yellow flowers at top of
leafy stem with pinnate leaves.
Flower 1" wide, petals 5,
round, spreading. Sepals 5,
round, with 5 short narrow
bracts between sepals. Stamens 20,
yellow. Ovaries superior, many on
central dome. Fruit cluster of achenes;
ovaries and achenes finely hairy. Stem
shrubby, brown shedding bark. Leaves 5–7
pinnate, narrow-oblong, leaflets $3/4$" long,
entire, often inrolled margins, finely hairy,
base of top 3 leaflets very close together;
alternate. Habitat: bogs, wet calcareous meadows.
FL: June–September.

SULFUR FIVE-FINGERS
Potentilla recta

ROSE FAMILY   ROSACEAE

Perennial, 2' tall, bearing 1-few
large yellow flowers on leafy
hairy stem with palmate
leaves. Flower 1" wide,
petals 5, heart-shaped.
Sepals 5, small, hairy,
with 5 narrow hairy bracts
below sepals. Stamens 20,
yellow. Superior ovaries on
central dome. Fruit striated achenes.
Leaves long-petioled, palmate, stipules,
alternate; leaflets 5–7, narrow oblong,
toothed. Flower-stem leaves, smaller. Habitat:
dry soil, roadsides; introduced. FL: July–August.
Rough F., *P. norvegica,* similar but only 3 leaflets, hairy,
smaller flowers, flat achenes. Habitat: disturbed areas, roadsides.

BARREN STRAWBERRY
Waldsteinia fragarioides
ROSE FAMILY   ROSACEAE

Perennial, 8" tall, strawberry-
like plant bearing several yellow
flowers on hairy leafless stalk
with adjacent 3-foliate leaves.
Flower $^1/_2$" wide, petals 5, round.
Sepals 5, narrow, short, united into
calyx-like cup. Stamens many,
yellow anthers. Ovary superior, 2–6
pistils, short styles. Fruit achenes
on persistent hairy calyx cup.
Leaflets 2" long, oval, hairy,
shallow-lobed, sharp- or
round-toothed, long hairy
petal, hairy underside, arise
from rhizome. Habitat: woods,
thickets. FL: April–June.

BIRDSFOOT-TREFOIL
Lotus corniculatus
PEA FAMILY   FABACEAE

Perennial, 1' tall, bearing
umbel of irregular yellow
flowers on long axillary
stalk on leafy stem with
pinnate leaves. Flower $1/2$"
wide, petals 5, large broad
upper petal forms upper lip,
lower keel lip of 2 fused petals, 2
lateral petals incurved. Calyx small 5-
pointed. Stamens 10, inside keel. Ovary
superior, 1 pistil. Fruit linear 4-angled
pod, persistent style. Umbel subtended by
entire bracts. Leaves petioled, pinnate with 5
sessile leaflets the lowest 2 close to stem, smooth, entire, no
stipule; alternate. Habitat: fields; introduced. FL: June–August.

BLACK MEDICK
Medicago lupulina
PEA FAMILY   FABACEAE

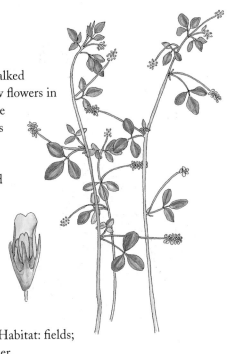

Annual, 1' tall, bearing long-stalked
umbels of small irregular yellow flowers in
axils of leafy stem with trifoliate
leaves. Flower $1/8$" long, 5 petals
are 2-lipped; upper lip petal is
large, oblong often notched; 2
laterals shorter than lip; 2 fused
petals form lower keel. Calyx
bell-shaped, 5-toothed.
Stamens 10. Ovary superior, 1
pistil, 1 style. Fruit curved
black pod, 1 seed. Stem often
prostrate. Leaves petioled,
alternate; leaflets $3/4$" oval,
round or notched tip, terminal
leaflet petioled, stipules small. Habitat: fields;
introduced. FL: June–September.

COMMON EVENING-PRIMROSE
Oenothera biennis

EVENING-PRIMROSE FAMILY   ONAGRACEAE

Biennial, 5' tall, bearing loose spike of large fragrant yellow flowers in axils of short bracts on leafy stem. Flower 1½" wide, petals 4, wide lobe with notch. Sepals 4, reflexed, narrow, pointed. Stamens 8, equal lengths, yellow. Ovary inferior, long style, X-shaped yellow stigma. Fruit 1" long, broadest near base, woolly, sessile. Flower stalk includes ovary and long floral tube. Flower opens late in day. Stem woolly. Lower leaves 7" long, upper shorter, lanceolate, slightly toothed, woolly underside, entire; alternate. Habitat: fields, roadsides. FL: June–September.

CROSS-SHAPED EVENING-PRIMROSE
Oenothera clelandii

EVENING-PRIMROSE FAMILY   ONAGRACEAE

Biennial, 3' tall, bearing several fragrant large yellow flowers on long stalk in axils of short bracts on leafy woolly stem. Flower 1" wide, corolla, 4 narrow ⅛" wide lobes. Calyx, longer than corolla, reflexed ⅛" wide lobes. Stamens 8, yellow. Ovary inferior, 1 pistil, long style, prominent X-shaped yellow stigma. Fruit ¾" long, sessile cylindrical capsule, not thickened at base. Stalk includes ovary and long floral tube. Flower opens late in day. Lower leaves 5" long, upper shorter, linear-oblong, taper to short petiole, entire. Habitat: sandy disturbed areas. FL: June–July.

SUNDROPS
Oenothera pilosella

EVENING-PRIMROSE FAMILY   ONAGRACEAE

Biennial, 2' tall, bearing a few large yellow flowers on long stalk in axils of bracts on leafy hairy stem. Flower 1" wide, petals 4, wide, notch at tip. Sepals 4, reflexed, pointed, long hairs. Stamens 8, orange. Ovary inferior, 1 pistil, long style, X-shaped yellow stigma. Fruit $^3/_4$" long, 4-angled pod, woolly, sessile. Stalk includes ovary and long floral tube. Bracts longer than ovary. Flower opens late in day. Stem red-dotted. Leaves 3" long, oblong-lanceolate, hairy underside, entire, taper at ends, nearly sessile; alternate. Habitat: moist fields, open woods. FL: May–July.

YELLOW WOOD-SORREL
Oxalis stricta

WOOD SORREL FAMILY   OXALIDACEAE

Perennial, 6" tall, bearing 1–4 yellow flowers on stalk in leaf axil on leafy hairy branched stem with clover-like leaves. Flower $^1/_2$" wide, petals 5, spreading. Sepals 5, shorter than petals. Stamens 10, both short and long. Ovary superior, 1 pistil, 5 styles. Fruit long capsule, erect, hairy. Leaves trifoliate on long petiole, each leaflet $^1/_2$" wide, heart-shaped, entire, hairy or smooth; alternate. Leaflets unfold in morning, fold at dusk. Habitat: disturbed areas, fields. FL: May–October.

GOLDEN ALEXANDERS
Zizia aurea

CARROT FAMILY   APIACEAE

Perennial, 2' tall, bearing
umbels of many fragrant
small yellow flowers on leafy
smooth stem with
trifoliate leaves.
Flower $1/8$" long, petals
5, tip curves inward,
oval. Sepals minute.
Stamens 5, short. Ovary
inferior, ridged, 2 persis-
tent styles. Fruit flat, rough, 5-ribbed.
Umbel with umbellets, subtended by bracts.
Flower stalks differ in length, but central flower sessile. Lower leaves
2× trifoliate, upper leaves 1× or irregularly so; alternate. Leaflets 2"
long, elliptical, toothed, pointed. Habitat: moist meadows, shores.
FL: May–June.

CLAMMY GROUND-CHERRY
Physalis heterophylla

NIGHTSHADE FAMILY   SOLANACEAE

Perennial, 2' tall, bearing large
nodding bell-shaped
yellow flower with
brown centers in
axil near top of
leafy hairy stem.
Flower $3/4$" wide,
corolla 5 spreading overlapping
partly united pointed lobes, brown
throat. Calyx 6-pointed, hairy, $1/2$
corolla length. Stamens 5, short, clustered at center of flower. Ovary superior.
Fruit yellow berry in inflated calyx. Stem
sticky, hairy, branched. Leaves 4" long, heart-shaped base, slightly
lobed, hairy, sticky, petioled; alternate. Habitat: woods, clearings.
Poisonous if eaten. FL: June–September.

BUTTER-AND-EGGS
Linaria vulgaris
FIGWORT FAMILY   SCROPHULARIACEAE

Perennial, 3' tall, bearing irregular yellow flowers in upper axils of leafy smooth stem. Flower 1¹/₂" wide, corolla 2-lipped, upper lip 2-lobed; lower lip 3-lobed dark yellow with orange palate. Base of lower lip forms dark yellow pointed spur. Calyx tube 5 short lobes. Stamens 4, below upper lip. Ovary superior, 1 pistil, style extends to throat. Fruit slender capsule, many rough seeds. Leaves 2" long, narrow, smooth, entire, taper to both ends, pale green; lower leaves opposite, upper alternate. Habitat: shaded roadsides, fields; introduced. FL: June–October.

WOOD-BETONY
Pedicularis canadensis
FIGWORT FAMILY   SCROPHULARIACEAE

Perennial, 1' tall, bearing dense terminal cluster of irregular tubular yellow hooded flowers on leafy stem with pinnately lobed leaves. Flower ³/₄" long, corolla 2-lipped, upper lip arched and incurved to form beak, lower lip 3-lobed and spreading downward. Calyx tube hairy, entire. Stamens 4, pairs of long and short, under lip. Ovary superior, 1 pistil, short style, globose stigma. Fruit long capsule. Bracts below flowers. Upper stem hairy. Leaves 5" long, hairy, oblong, pinnately cleft into lobes, petioled; opposite-alternate. Habitat: dry woods, thickets. FL: April–June.

COMMON MULLEIN
Verbascum thapsus
FIGWORT FAMILY   SCROPHULARIACEAE

Biennial, 6' tall, bearing dense club-shaped terminal cluster of large slightly irregular yellow flowers on leafy hairy stem. Flower 1" wide, corolla tube 5 unequal, large lobes. Calyx 5-lobed, hairy. Stamens 5, upper 3 hairy and lower 2 smooth, protrude above throat. Ovary superior, 1 pistil, long style, extend above corolla. Fruit capsule, brown pitted seeds. Stem woolly, may be branched near top. Basal leaves 12" long, elliptic, hairy, petioled, felt-like appearance. Upper leaves elongated, sessile, wings extend onto stem; alternate. Habitat: fields, roadsides; introduced. FL: June–September.

NAKED BLADDERWORT
Utricularia cornuta
BLADDERWORT FAMILY   LENTIBULARIACEAE

Perennial, 6" tall, bearing several large spurred irregular yellow flowers on leafless stalk. Flower $1/2$" wide, corolla 2-lipped, upper lip erect, shorter than lower lip. Lower lip broad, somewhat 3-lobed with elevated palate with spreading margin, and tube forming long pendant spur nearly length of lip. Calyx, 2-parted, small. Stamens 2. Ovary superior, 1 pistil. Fruit capsule. Flower subtended by several small bracts. Stalk bears several scale leaves. Stem, roots and leaves subterranean. Carnivorous, bladders on roots trap and digest organisms. Habitat: wet sandy shores. FL: June–August.

COMMON BLADDERWORT
Utricularia vulgaris

BLADDERWORT FAMILY   LENTIBULARIACEAE

Perennial aquatic, 2' long, bearing several large irregular yellow flowers on stalk attached to stem with under-water leaves. Flower $^3/_4$" wide, corolla 2-lipped. Upper lip hood-like; lower lip broad, 3-lobed with elevated palate and spur nearly length of lip. Calyx 2-lipped. Stamens 2. Ovary superior, 1 pistil. Fruit capsule. Flower subtended by bract; scales on stalk. Stem floating or submerged. Leaves $1^1/_4$" long, thread-like. Carnivorous, many small oval bladders on leaves trap and digest aquatic organisms. Habitat: quiet inlets, ponds. FL: June–August.

NORTHERN BUSH-HONEYSUCKLE
Diervilla lonicera

HONEYSUCKLE FAMILY   CAPRIFOLIACEAE

Perennial, 2' tall, bearing clusters of 2–3 fragrant irregular funnel-shaped yellow flowers at top of somewhat woody leafy stem. Flower $^3/_4$" long, corolla 5 spreading lobes, 3 pointed upward, 2 curled downward. Calyx 5-pointed, small. Stamens 5, pale orange anthers. Ovary inferior, long style, green stigma above anthers. Fruit capsule. Leaves 4" long, oval, pointed, toothed, smooth, short-petioled; opposite. The only honeysuckle with toothed leaves. Habitat: rich maple woods. FL: June–August.

Variety, *hypomalaca*, has toothless leaves, velvety lower leaf surface.

HAIRY HONEYSUCKLE
Lonicera hirsuta

HONEYSUCKLE FAMILY   CAPRIFOLIACEAE

Perennial vine, 10' long, bearing 1–3 whorls of irregular tubular yellow flowers at top of stem above perfoliate leaves. Flower $3/4$" wide, hairy corolla tube cleft into 2 lips, the upper lip narrow and lower lip wide with 4 tiny lobes at tip, tube base swollen on one side. Stamens 5, yellow, extend above corolla. Ovary inferior, style extends above corolla. Fruit berry. Stem woolly, glandular. Leaves 5" long, finely hairy, oval, pointed, entire, petioled, upper pairs perfoliate under flowers; opposite. Habitat: moist woods. FL: June–July.

NODDING BUR-MARIGOLD
Bidens cernua

DAISY FAMILY   ASTERACEAE

Annual, 3' tall, bearing large yellow flower-head with pointed broad yellow rays and light brown center on leafy hairy stem. Head $1^1/_2$" wide, 6–10 ray, and disk flowers. Disk flower corolla 5-lobed. Calyx pappus of 4 barbs. Stamens 5. Ovary inferior, 1 pistil, style branches appendaged. Fruit narrow achene, 4 barbs, recurved hairs; stick to fur, clothing. Involucral bracts 5–10, yellow tips, equal rays. Head nods with age. Leaves 7" long, narrow, toothed, smooth, pointed, sessile; opposite. Habitat: wet woods, ditches. FL: August–October.

STRAWSTEM BEGGAR-TICKS
Bidens comosa
DAISY FAMILY   ASTERACEAE

Annual, 3' tall, bearing yellow-brown flower-heads with long leafy bracts on stalks in axils of smooth leafy stem. Head $^3/_8$" wide, disk flowers only. Corolla 4-lobed. Calyx pappus 3 barbs. Stamen anthers not above corolla. Ovary inferior, 1 pistil, style branches with hairy appendages. Fruit flat brown hairy achene, 3 barbs with 1 shorter, recurved hairs, ridged; stick to fur, clothing. Involucral bracts 4–8, long, green, smooth. Stem branched near top. Leaves 4" long, taper at both ends, narrow, coarsely toothed; opposite. Habitat: moist woods, ditches. FL: July–September.

BEGGAR-TICKS
Bidens frondosa
DAISY FAMILY   ASTERACEAE

Annual, 3' tall, bearing yellow-brown flower-heads with hairy leafy bracts on long stalks in axils of smooth leafy stem. Head $^3/_8$" wide, disk flowers only; a few ray rarely present. Corolla 5-lobed. Calyx pappus 2 elongated barbs. Stamen anthers above corolla. Ovary inferior, 1 pistil, style branches with hairy appendages. Fruit flat hairy achene, 2 barbs, recurved hairs; stick to fur, clothing. Involucral bracts 8–10, long, leaf-like, hairy. Leaves 4" long, pinnate, 3–5 leaflets, petioled, toothed, hairy underside; opposite. Habitat: wet woods, ditches. FL: July–September.

LONGSTALK-TICKSEED
Coreopsis lanceolata

DAISY FAMILY ASTERACEAE

Perennial, 2' tall, bearing large flower-head
with notched yellow rays and yellow
center on nearly leafless smooth stalk.
Head 2" wide, 8 notched rays with
long veins; many disk flowers. Disk
flower corolla tubular, 5-lobed. Calyx
pappus of short teeth. Stamens 5, short.
Ovary inferior, 1 pistil, style hairy.
Fruit flat winged achene, pappus.
Long pointed bracts among disk
flowers. Involucral bracts 2 rings,
outer bracts oval, short and inner long,
narrow. Lower leaves petioled,
lanceolate, smooth, entire, opposite;
upper small, sessile. Habitat: dry sandy areas. FL: May–June.

SNEEZEWEED
Helenium autumnale

DAISY FAMILY ASTERACEAE

Perennial, 4' tall, bearing several
globose flower-heads with drooping
yellow center at top of winged leafy stem.
Head 2" wide, both disk and ray flowers. Disk
flower corolla lobes hairy. Ray flower corolla
fan-like, drooping, 3-lobed. Calyx pappus of
short pointed scales. Ovary inferior, 1
pistil, style branches flat.
Fruit fuzzy achene, pappus.
Involucral bracts loose,
reflexed. Leaves 6" long,
lanceolate, widely spaced teeth,
taper to ends, red edge of leaf
extends onto and down stem; alternate. Habitat: swamps, wet meadows.
FL: August–November.

SAWTOOTH SUNFLOWER
Helianthus grosseserratus
DAISY FAMILY   ASTERACEAE

Perennial, 6' tall, bearing flower-head with yellow rays and yellow-brown center on leafy stem. Head $3^{1}/_{2}$" wide, 10–20 ray, and disk flowers. Disk flower corolla tubular. Calyx pappus of 2 awns. Stamen 5. Ovary inferior, style tips hairy. Fruit flat achene, pappus none. Scaly bracts among disk flowers. Involucral bracts narrow, pointed, edge fringed. Appressed fine hairs near head, smooth below. Leaves 8" long, lanceolate, petiole often winged, lower toothed, appressed hairs on top, short hairs on underside; opposite below, alternate above. Habitat: moist areas, meadows. FL: July–September.

DIVARICATE SUNFLOWER
Helianthus divaricatus
DAISY FAMILY   ASTERACEAE

Perennial, 6' tall, bearing 1-few large flower-heads with yellow rays and center disk on smooth stem with opposite leaves. Head 3" wide, 8–15 ray, and disk flowers. Stamens 5. Ovary inferior, 1 pistil, style tips hairy. Fruit smooth ridged black achene. Scaly bracts among disk flowers. Involucral bracts narrow, loose, not overlapping, reflexed tips. Leaves 6" long, widest near rounded base, petiole length $^{1}/_{4}$" or less, taper to long pointed tip, finely hairy underside, toothed. Habitat: dry open woods. FL: July–September.

MAXIMILIAN SUNFLOWER
Helianthus maximilianii

Daisy family   Asteraceae

Perennial, 4' tall, bearing large
flower-head with yellow rays and
yellow-brown center on stem with
partly folded leaves. Head 3" wide, 10–20 ray,
and disk flowers. Disk flower corolla tubular.
Stamens 5. Ovary inferior, 1 pistil, style tips
hairy. Fruit flat plump achene.
Scaly bracts among disk
flowers. Involucral bracts
pointed. Stem finely hairy
below head. Leaves 6" long, narrow,
folded inward along midrib, tip curved
downward, rough to touch, gray-green velvety,
sessile or winged petiole, hardly toothed,
often widest below middle; alternate above.
Habitat: dry open woods. FL: July–October.

HAIRY SUNFLOWER
Helianthus hirsutus

Daisy family   Asteraceae

Perennial, 6' tall, bearing large flower-
head with yellow rays and brown
center on leafy stem with opposite
petioled leaves. Head 3" wide, 10–14
ray, and tubular disk flowers. Pappus 2
awns, drop early. Ovary inferior, style tips
hairy. Fruit flat achene, pappus none.
Involucral bracts finely hairy on margin and
vein, reflexed tips. Stem hairs spreading
and appressed, hairy base often enlarged.
Leaves 5" long, hairy, $1/2$" long
petiole slightly winged,
widest near round base, 3-
veined, veins hairy on underside,
pointed: opposite. Habitat: dry woods, open
areas. FL: July–October.

SMOOTH HAWKWEED
Hieracium floribundum
DAISY FAMILY   ASTERACEAE

Perennial, 4' tall, bearing several
yellow flower-heads each on long
stalk at top of tall slightly hairy
stem above basal leaves. Head 1"
wide, ray flowers only.
Corolla square-tipped,
5-notched. Calyx dull
pappus bristles. Stamens 5,
encircle style. Ovary inferior, 1
pistil, lobed style. Fruit tubular achene,
pappus. Involucral bracts fuzzy. Bracts
below long stalk of heads. Inflorescence stem
with long hairs; exudes milky sap. Basal
leaves 6" long, widest above middle,
linear, hairs along margin. Habitat:
meadows, roadsides; introduced.
FL: June–August.

DWARF DANDELION
Krigia virginica
DAISY FAMILY   ASTERACEAE

Annual, 8" tall, bearing one yellow
flower-head at top of leafless stalk
above rosette of leaves. Head $^3/_4$"
wide, ray flowers only. Corolla
square-tipped, 5-notched. Calyx
pappus 5-scales and up to 10 bristles.
Stamens 5. Ovary inferior, 1 pistil, forked
style. Fruit nerved achene, pappus. Involucral bracts 9–18, reflexed on older flower.
Flower-head stalk smooth or glandular near
head, exudes milky sap.
Basal leaves 5" long,
oblong-elongate, shallowly
toothed, narrow to base.
Habitat: sandy areas. FL:
May–September.

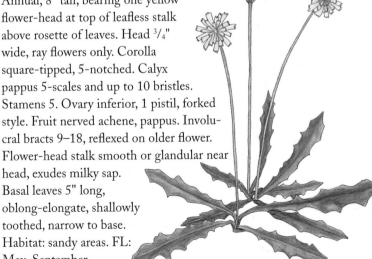

TALL LETTUCE
Lactuca canadensis

DAISY FAMILY   ASTERACEAE

Biennial, 8' tall, bearing many
small yellow flower-heads on short
branches on smooth stem with milky
sap. Heads $^3/_8$" wide, 12 or more ray
flowers only. Corolla edges inrolled, tip
notched. Calyx pappus of bristles.
Stamens 5. Ovary inferior, long
persistent style. Fruit flat
smooth brown achene,
pappus on raised beak.
Involucral bracts green.
Leaves 1' long, deeply
pinnate, linear lobes,
sessile base winged, lobes clasp stem, veins
hairy on underside. Habitat: moist rich woods. FL: July–August.
Tall Blue L., *L. biennis,* similar but flowers violet, pappus at tip of
ribbed achene (below). Habitat: open areas in moist woods.

NIPPLEWORT
Lapsana communis

DAISY FAMILY   ASTERACEAE

Annual, 3' tall, bearing loose group of
yellow flower-heads in upper axils of
leafy stem. Head $^1/_2$" wide, 8–15 ray
flowers only. Ray corolla 5-notched.
Calyx pappus none. Stamens 5, encircle
style. Ovary inferior, 1 pistil. Fruit
curved achene, pappus none.
Involucral bracts smooth,
pointed, with pale bloom.
Stem slender, hairy below
and smooth above,
branched, exudes milky sap. Lower leaves 4"
long, oval, coarsely toothed, petioled, wing-like
lobes along petiole. Upper leaves small, lanceolate, sessile; alternate.
Habitat: roadsides, disturbed areas; introduced. FL: July–August.

GLOBULAR CONEFLOWER
Ratibida pinnata
<small>DAISY FAMILY   ASTERACEAE</small>

Perennial, 4' tall, bearing scented flower-head with drooping yellow rays and gray-brown cone on leafy stalk. Head 2$^1/_2$" wide, 6–8 ray, and many disk flowers. Disk flower corolla 5-lobed. Calyx pappus none. Stamens 5. Ovary inferior, 1 pistil, style tips flat, appendaged. Fruit 2-angled smooth achene. Cone cylindric, shorter than ray corolla. Bracts among disk flowers. Involucral bracts in ring, scaly. Leaves 5" long, pinnate lobes, lanceolate segments, finely hairy, entire; alternate; stalk leaves smaller, lanceolate, finely hairy. Habitat: dry wood openings. FL: June–August.

EASTERN CONEFLOWER
Rudbeckia fulgida
<small>DAISY FAMILY   ASTERACEAE</small>

Perennial, 3' tall, bearing large flower-head with yellow-orange rays and brown center on leafy hairy stem. Head 3" wide, 10–20 ray, and disk flowers. Disk corolla 5-lobed. Calyx pappus of slight crown. Stamens 5. Ovary inferior, 1 pistil, style hairy. Fruit smooth angular achene, pappus. Smooth blunt bracts among disk flowers. Involucral bracts green, hairy, black-tipped. Stem red-streaked. Leaves: lower 6" long, lanceolate, hairy, finely toothed, petioled; upper similar but smaller, sessile, alternate. Habitat: rocky woods openings. FL: July–October.

CUTLEAF-CONEFLOWER
Rudbeckia laciniata
DAISY FAMILY   ASTERACEAE

Perennial, 5' tall, bearing flower-head with pointed drooping yellow rays and green-yellow center on smooth stem with lobed leaves. Head 3" wide, 10–20 rays, and disk flowers. Disk flower corolla 5-lobed. Calyx pappus a short crown. Stamens 5. Ovary inferior, 1 pistil, style short, blunt. Fruit smooth 4-angled achene, pappus. Blunt sticky bracts among disk flowers. Involucral bracts smooth, green, reflexed. Leaves long, petioled, 3- or pinnately lobed, toothed, finely hairy underside; alternate. Habitat: moist woods, stream edges. FL: July–September.

BLACK-EYED SUSAN
Rudbeckia hirta
DAISY FAMILY   ASTERACEAE

Perennial, 3' tall, bearing large flower-head with yellow rays and brown center on leafy hairy stem. Head 3" wide, 8–12 ray and many disk flowers. Disk flower corolla, 5-lobed. Calyx pappus none. Stamens 5, encircle style. Ovary inferior, 1 pistil, style hairy. Fruit smooth angular achene, pappus none. Hairy spatulate bracts among disk flowers. Involucral bracts elongate, hairy, spreading. Lower leaves 6" long, oval-lanceolate, rough, hairy, prominently 3-veined, winged petiole. Leaves on stem similar but ovate, smaller, sessile; alternate. Habitat: dry to moist open woods, fields. FL: June–October.

CANADA GOLDENROD
Solidago canadensis

Perennial, 5' tall, bearing many small yellow flower-heads in broad terminal plume on leafy stem. Head $1/8$" wide, 9–15 ray, and several disk flowers. Calyx pappus of bristles. Stamens 5. Ovary inferior, forked style hairy. Fruit ribbed slightly hairy achene, pappus. Involucral bracts yellowish, pointed, in rings. Stem hairy above but smooth below middle. Lower leaves 5" long, thin, narrow-lanceolate, 3-veined, toothed, sessile; alternate. Upper leaves entire, crowded under plume. Gall Fly forms round galls on stem. Habitat: open meadows. FL: July–September.

ZIGZAG GOLDENROD
Solidago flexicaulis

DAISY FAMILY   ASTERACEAE

Perennial, 2' tall, bearing several yellow flower-heads in upper axils on arched zigzag leafy stem. Head $3/16$" wide on short stalk, 3–4 ray, and 5–7 disk flowers with spreading corolla lobes. Calyx pappus of bristles. Ovary inferior, 1 pistil, forked style. Fruit minutely hairy achene, pappus bristles. Involucral bracts blunt-tipped, smooth. Stem smooth, edges angled. Leaves 3" long, ovate, taper to winged petiole, finely toothed, pointed; alternate. Habitat: rich shaded woods; colonial. FL: July–September.

HAIRY GOLDENROD
Solidago hispida

DAISY FAMILY   ASTERACEAE

Perennial, 2' tall, bearing several yellow
flower-heads in upper axils of bracts
forming spike inflorescence on leafy finely
hairy stem with basal leaves. Head $^3/_{16}$"
wide, 7–12 ray, and fewer disk flowers with
spreading corolla lobes. Calyx pappus of
bristles. Stamens 5. Ovary inferior, 1 pistil,
forked style. Fruit smooth achene, pappus.
Involucral bracts narrow, many, overlapping. Basal leaves 6" long × 1$^1/_2$" wide,
hairy, long-petioled, wavy margin, blunt
tip. Stem leaves small, linear, hairy,
often sessile, entire; alternate. Habitat:
rocky shores. FL: July–October.

EARLY GOLDENROD
Solidago juncea

DAISY FAMILY   ASTERACEAE

Perennial, 3' tall, bearing many small
yellow flower-heads in broad
terminal plume with recurved
branches on smooth leafy stem
with basal leaves. Head $^1/_8$" wide,
7–12 ray, and 4–several disk
flowers with spreading
corolla lobes. Calyx pappus
of bristles. Stamens 5. Ovary
inferior, 1 pistil, forked style.
Fruit woolly achene, pappus.
Involucral bracts smooth, pointed.
Basal leaves 12" long × 2" wide,
long-petioled, toothed, pointed,
smooth; upper leaves 2" long,
smaller, lanceolate, smooth, entire, sessile,
alternate. Habitat: dry open areas; first golden-
rod to appear. FL: June–September.

GRAY GOLDENROD
Solidago nemoralis
Daisy family   Asteraceae

Perennial, 2' tall, bearing many small yellow
flower-heads on upper surface of nodding wand
inflorescence on leafy hairy stem with basal leaves.
Head $^1/_8$" wide, 5–8 ray, and fewer disk flowers with
spreading corolla lobes. Calyx pappus of bristles.
Stamens 5. Ovary inferior, 1 pistil, forked style.
Fruit woolly achene, pappus. Stem reddish-gray.
Basal leaves 8" long × 1$^1/_2$" wide, lanceolate,
long-petioled, 3-veined, hairy, toothed,
pointed. Stem leaves small, gray-green, downy,
narrow, lowermost toothed; alternate. Habitat:
dry woods, meadows. FL: August–October.

OHIO GOLDENROD
Solidago ohioensis
Daisy family   Asteraceae

Perennial, 3' tall, bearing yellow flower-heads in
flat-topped inflorescence on leafy stem
with basal leaves. Head $^1/_8$" wide,
5–8 ray, and disk flowers
(below). Calyx pappus of
bristles. Stamens 5. Ovary
inferior, pistil with forked
style. Fruit 3–5 angled
achene, pappus. Involucral
bracts smooth, blunt-tipped.
Basal leaves 8" long, pointed,
long-petioled; upper leaves 2" long,
narrow but over $^3/_8$" wide, smooth,
entire, sessile, alternate. Habitat: moist
meadows, bay edges. *Special Concern.* FL:
July–September.
Flat-topped G., *Euthamia graminifolia,*
very narrow $^1/_8$–$^3/_8$" wide glandular
leaves. Habitat: sandy lakeshores.

DUNE GOLDENROD
Solidago simplex, var. *gillmanii*

DAISY FAMILY   ASTERACEAE

Perennial, 4' tall, bearing spike of large yellow
flower-heads on leafy smooth stem with basal
leaves. Head ¼" wide, 8–10 ray and fewer disk
flowers. Disk flower with spreading lobes
(above). Calyx pappus of bristles.
Ovary inferior, forked style. Fruit
woolly achene, pappus. Involucral
bracts narrow, overlapping, pointed;
glandular hairs among disk flowers.
Basal leaves 12" long × 1¼" wide,
long-petioled, wavy margin; upper
leaves 4" long or shorter, toothed to
entire, smooth, sessile; alternate.
Threatened. Habitat: lakeside sand
dunes. FL: July–October.

COMMON SOW-THISTLE
Sonchus oleraceus

DAISY FAMILY   ASTERACEAE

Annual, 5' tall, bearing yellow
flower-heads of ray flowers at
top of smooth stem
with clasping leaves
and milky sap. Head 1" wide, 80 or
more ray flowers only. Corolla
square-tipped, 5-notched. Calyx
pappus of bristles.
Stamens 5. Ovary
inferior, 1 pistil, forked
style. Fruit flat achene, each side 3-
ribbed, pappus. Involucral bracts
overlapping, green. Leaves 9" long, elongate,
scarcely prickly, leaf base round-lobed, toothed.
Habitat: disturbed areas; introduced. FL: June–October.
Prickly S-t., *S. asper* (right), similar but leaves prickly-edged, base very
round-lobed, clasping stem. Habitat: disturbed areas; introduced.

COMMON TANSY
Tanacetum vulgare

Dᴀɪsʏ ꜰᴀᴍɪʟʏ   Aꜱᴛᴇʀᴀᴄᴇᴀᴇ

Perennial, 3' tall, bearing flat-topped cluster of yellow flower-heads at top of aromatic smooth stem with pinnate leaves. Head ¹/₂" wide, 20 or more disk flowers only. Corolla tubular, 5-toothed. Calyx pappus a small crown. Stamens 5, encircle style. Ovary inferior, 1 pistil, style lobes short, flat. Fruit 5-ribbed glandular achene, pappus absent. Flowers perfect; a few marginal flowers may form small ray-like petals. Leaves 8" long, pinnately divided into tiny narrow leaflets, toothed; alternate. Habitat: roadsides, fields; introduced. FL: July–September.

DANDELION
Taraxacum officinale

Dᴀɪsʏ ꜰᴀᴍɪʟʏ   Aꜱᴛᴇʀᴀᴄᴇᴀᴇ

Perennial, 1' tall, bearing yellow flower-head at top of leafless stalk above rosette of leaves. Head 1¹/₄" wide, ray flowers only. Ray corolla square-tipped. Calyx pappus of bristles. Stamens 5, encircle style. Ovary inferior, 1 pistil. Fruit achene, pappus. Involucral bracts 13–20 in 2 rings, outer shorter than inner, and reflexed. Flower-head stalk finely hairy above, smooth below, exudes milky sap. Leaves basal 8" long, elongate, pinnately lobed with tips of lobes pointing toward base, narrows toward base. Habitat: disturbed areas; introduced. FL: April–November.

FISTULOUS GOAT'S-BEARD
Tragopogon dubius

DAISY FAMILY   ASTERACEAE

Perennial, 2' tall, bearing typically one
large yellow flower-head at top of leafy
stem with grass-like leaves and exudes
milky sap. Head 1¼" wide, ray flowers
only. Ray corolla square-tipped. Calyx
pappus of plumose bristles. Stamens 5,
encircle style. Ovary inferior, 1 pistil.
Fruit long-beaked achene, pappus.
Involucral bracts longer than rays. Stem
diameter broadens at very base of
flower-head. Leaves 1' long × ⅝" wide,
narrow, keeled, smooth, clasp stem.
Habitat: roadsides, disturbed areas;
introduced. FL: May–July.

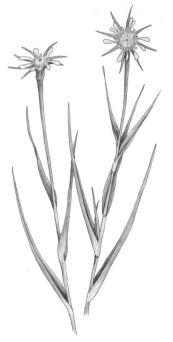

SHOWY GOAT'S-BEARD
Tragopogon pratensis

DAISY FAMILY   ASTERACEAE

Perennial, 2' tall, bearing typically
one large yellow flower-head at
top of leafy stem with grass-like
leaves and exudes milky sap. Head
1¼" wide, ray flowers only. Ray
corolla square-tipped. Calyx pappus
plumose bristles. Stamens 5. Ovary
inferior, 1 pistil. Fruit angular
achene, pappus. Involucral bracts
only as long as rays. Stem diameter
uniform to very base of flower-head.
Leaves 1' long × ¾" wide, keeled,
smooth, clasp stem. Habitat: roadsides,
disturbed areas; introduced. FL: May–August.

WATER STAR-GRASS
Zosterella dubia

<small>WATER-HYACINTH FAMILY</small>   <small>PONTEDERIACEAE</small>

Perennial aquatic, 2" tall, bearing solitary
yellow flower on leafless stalk from basal
cluster of elongated leaves. Flower ³/₄"
wide, slender perianth tube with 6
spreading tepals, narrow, blunt-tipped.
Stamens 3, dilated filaments, yellow anthers.
Ovary superior 3 chambered, elongated
style, dark forked
stigma. Fruit narrow
oval capsule. Stem
creeping, often
submerged. Leaves
grass-like, linear, pointed or
blunt-tipped. Habitat: sand or
mudflats. Flower stalk extends to surface; rare. FL: August.

BEAD-LILY
Clintonia borealis

<small>LILY FAMILY</small>   <small>LILIACEAE</small>

Perennial, 16" tall, bearing 3–6 bell-
shaped yellow-green flowers on long
smooth stalk rising between basal set of
wide shiny leaves. Flower 1" long,
tepals 6, nodding.
Stamens 6, as long
as tepals, yellow
anthers. Ovary
superior, 3-celled,
short stigma. Fruit globose
shiny blue ¹/₂" berry, several
brown seeds. Leaves 10" long,
broad, thick, elliptical, entire,
taper to reddish base. Leaf veins
appear parallel. Habitat: moist
woods. FL: May–August.

TROUT-LILY
Erythronium americanum

LILY FAMILY   LILIACEAE

Perennial, 10" tall, bearing
single large nodding yellow
flower at tip of leafless
stalk with 2 en-
sheathing leaves.
Flower 2"
wide, perianth
6 separate
tepals, yellow inner surface,
green-purple outer surface, base
dotted with brown spots, all
tepals curved backward. Anthers
6, long, brownish. Ovary
superior, 3-celled, 1 pistil, long
style. Fruit capsule, many brown
seeds. Leaves 8" long, smooth,
shiny, taper at ends, mottled with brown or purple spots, surround
floral stalk. Habitat: rich moist woods. FL: April–May.

PERFOLIATE BELLWORT
Uvularia grandiflora

LILY FAMILY   LILIACEAE

Perennial, 2' tall, bearing
single large nodding bell-
shaped yellow flower on
short axillary stalks on forked
stem with perfoliate leaves.
Flower 2" long, perianth 6
tepals, yellow, glandular inner
surface. Stamens 6, surround 3-
lobed superior ovary with 3 styles-
stigmas. Fruit lobed capsule, ariled brown
wrinkled seeds. Stem upright with arching
tips, forked at mid-height, clasping leaves
below fork. Leaves 3" long, ovate, stem pierces leaf, entire, pointed,
downy underside. Habitat: moist limestone woods. FL: April–June.

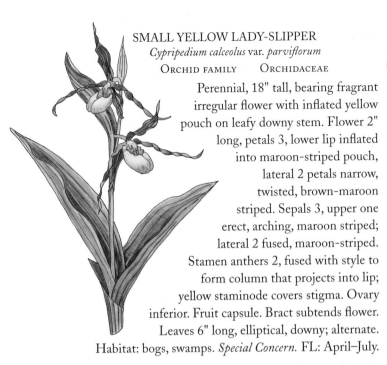

SMALL YELLOW LADY-SLIPPER
Cypripedium calceolus var. *parviflorum*
Orchid family   Orchidaceae

Perennial, 18" tall, bearing fragrant irregular flower with inflated yellow pouch on leafy downy stem. Flower 2" long, petals 3, lower lip inflated into maroon-striped pouch, lateral 2 petals narrow, twisted, brown-maroon striped. Sepals 3, upper one erect, arching, maroon striped; lateral 2 fused, maroon-striped. Stamen anthers 2, fused with style to form column that projects into lip; yellow staminode covers stigma. Ovary inferior. Fruit capsule. Bract subtends flower. Leaves 6" long, elliptical, downy; alternate. Habitat: bogs, swamps. *Special Concern.* FL: April–July.

LARGE YELLOW LADY-SLIPPER
Cypripedium calceolus var. *pubescens*
Orchid family   Orchidaceae

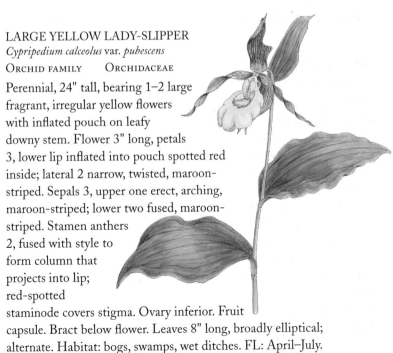

Perennial, 24" tall, bearing 1–2 large fragrant, irregular yellow flowers with inflated pouch on leafy downy stem. Flower 3" long, petals 3, lower lip inflated into pouch spotted red inside; lateral 2 narrow, twisted, maroon-striped. Sepals 3, upper one erect, arching, maroon-striped; lower two fused, maroon-striped. Stamen anthers 2, fused with style to form column that projects into lip; red-spotted staminode covers stigma. Ovary inferior. Fruit capsule. Bract below flower. Leaves 8" long, broadly elliptical; alternate. Habitat: bogs, swamps, wet ditches. FL: April–July.

Red to
Orange

FLOWERS

WILD GINGER
Asarum canadense

BIRTHWORT FAMILY   ARISTOLOCHIACEAE

Perennial, 8" tall, bearing at
ground level a single maroon
flower in crotch between 2
heart-shaped finely hairy
leaves. Flower $1^1/_2$" wide,
petals none. Calyx cup-
shaped, hairy, with 3 sharply
pointed maroon lobes, often
reflexed; outer cup surface creamy-white.
Stamens 6, thick, maroon anthers. Ovary
inferior, 6 chambers, 6 cohering styles
form 6-lobed stigma. Fruit capsule, large
oval ariled seeds. Stem rhizome forms 2
large petioled 6" wide heart-shaped entire
leaves. Crushed stem, leaves have ginger
odor. Habitat: rich woods, colonial. FL: April-May.

CANADA-COLUMBINE
Aquilegia canadensis

BUTTERCUP FAMILY   RANUNCULACEAE

Perennial, 3' tall, bearing large nodding
red and yellow flowers with
upward spurred petals on stalks
from axils on leafy smooth stem.
Flower 2" long, 5 upright red
spurs each with yellow petal lobe
facing downward. Sepals 5, red,
between petals, project downward.
Stamens many, long. Ovary superior,
5 pistils, long thin style. Fruit beaked
dry pod, many black seeds. Flower
subtended with bracts. Stem smooth,
delicate. Leaves 6" wide, long-petioled,
divided 2–3× lobed, leaflets spatulate, tip lobe
petioled; alternate. Habitat: open dry soil, rocky
slopes. FL: April–July.

RED SORREL
Rumex acetosella

SMARTWEED FAMILY   POLYGONACEAE

Perennial, 1' tall, bearing numerous
tiny red flowers in narrow loose
spikes on red stem with green
arrowhead-shaped leaves. Plants
with male or female flowers, $^{1}/_{16}$"
wide. Male: nodding, perianth 6
tepals, 6 large stamens
(shown). Female: perianth 6
tepals, inner 3 larger than
outer, ovary superior, round-
ish, 3 lobed styles hang down
around ovary, tiny staminodes.
Fruit winged achene. Leaves $1^{1}/_{2}$"
long, basal lobes pointing outward or
upward, smooth, entire. Habitat: sandy
open areas; introduced. FL: June–August.

MARSH ST. JOHN'S-WORT
Triadenum virginicum

MANGOSTEEN FAMILY   CLUSIACEAE

Perennial, 2' tall, bearing small
clusters of normally closed
maroon flowers at top of
stem with red-dotted
leaves. Flower $^{3}/_{8}$" long,
petals 5, appear as bud.
Sepals 5, short, pointed. Stamens
9, in groups of 3, basal glands.
Ovary superior, 1 pistil, 3-celled, 3
styles. Fruit cylindrical capsule, taper
to persistent styles. Stem unbranched,
smooth. Leaves $2^{1}/_{4}$" long, linear, red
translucent spots (glands) on top, sessile,
entire; opposite. Habitat: bogs, marshes.
FL: July–August.

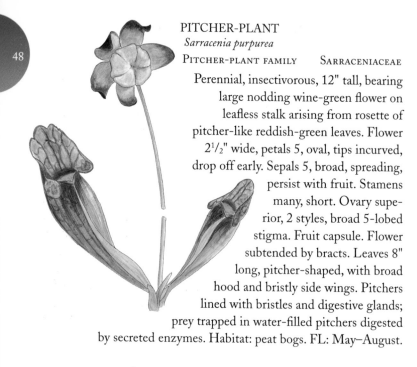

PITCHER-PLANT
Sarracenia purpurea
PITCHER-PLANT FAMILY   SARRACENIACEAE

Perennial, insectivorous, 12" tall, bearing large nodding wine-green flower on leafless stalk arising from rosette of pitcher-like reddish-green leaves. Flower 2$^1/_2$" wide, petals 5, oval, tips incurved, drop off early. Sepals 5, broad, spreading, persist with fruit. Stamens many, short. Ovary superior, 2 styles, broad 5-lobed stigma. Fruit capsule. Flower subtended by bracts. Leaves 8" long, pitcher-shaped, with broad hood and bristly side wings. Pitchers lined with bristles and digestive glands; prey trapped in water-filled pitchers digested by secreted enzymes. Habitat: peat bogs. FL: May–August.

LOWBUSH-BLUEBERRY
Vaccinium angustifolium
HEATH FAMILY   ERICACEAE

Perennial, 30" tall, bearing small clusters of nodding red urn-shaped flowers in upper axils of bushy leafy stems. Flower $^3/_{16}$" long, corolla tube 5 short teeth. Calyx 5-toothed, short, persistent. Stamens 10, short, within tube. Ovary inferior, slender style. Fruit juicy blue berry. Leaves 2" long, elliptic, smooth, some hairs on midrib on underside, small marginal teeth, nearly sessile; alternate. Habitat: sandy woods near bogs. FL: May–June.

MARSH-POTENTILLA
Potentilla palustris

ROSE FAMILY   ROSACEAE

Perennial, 2' tall, bearing small cluster of maroon flowers at top of leafy finely hairy stiff stem with pinnate leaves. Flower $3/4$" wide, petals 5, ovate, pointed. Sepals 5, small, finely hairy. Stamens many, short. Ovaries, superior, many on central dome, style lateral. Fruit achenes, lateral style. Bract below flower stalk. Leaves 5-pinnate, long-petioled, upper leaves sessile; alternate. Leaflets 3" long, narrow, linear, toothed, upper 3 adjacent at their bases. Habitat: swamps, stream banks. FL: June–August.

TOUCH-ME-NOT
Impatiens capensis

TOUCH-ME-NOT FAMILY   BALSAMINACEAE

Annual, 5' tall, bearing dangling irregular cone-shaped orange brown-spotted flowers on long stalks in axils of leafy smooth stem. Flower 1" long, corolla 5-parted, upper 1 broad, each lateral narrow and curl inward, lower 2 are spatulate. Calyx 3-lobed, large lobe forms inflated sac with slender curved spur; other 2 lobes are green, tiny. Stamens 5, united anthers. Ovary superior. Fruit capsule 1" long, explodes to release seeds. Leaves 3" long, oval, thin, smooth, toothed, pale green, downy underside, petioled; alternate. Habitat: wet woods, shores. FL: July–September.

HOUND'S TONGUE
Cynoglossum officinale

BORAGE FAMILY   BORAGINACEAE

Biennial, 4' tall, bearing bell-shaped
wine flowers along forked
coiled branches at top of
leafy hairy stem. Flower $^3/_8$"
wide, corolla 5 flared lobes.
Calyx 5-lobed, hairy.
Stamens 5, dark wine, short. Ovary
superior, short stigma-style. Fruit 4
nutlets, outer surface of each nutlet
covered with hooked
hairs especially along
edges. Stem with coiled tips when young, hairy.
Lower leaves 8" long, finely hairy, lanceolate,
entire, pointed, taper to petiole. Upper leaves
progressively smaller, nearly sessile; alternate. Habitat: fields, open
woods; introduced. FL: May–August.

PAINTED CUP
Castilleja coccinea

FIGWORT FAMILY   SCROPHULARIACEAE

Biennial, 2' tall, bearing terminal
cluster of tubular red-orange and
green-based bracts at top of leafy
stem. Flower corolla tiny, 2-lipped,
and calyx covered by bract. Stamens
4, tiny, under corolla lip. Superior ovary
light green with 1 pistil, elongated style
projects above top of red bracts. Fruit long
capsule, many seeds. Stem minutely hairy,
unbranched. Rosette leaves 3" long, ovate,
entire, hairy. Stem leaves hairy, deeply 3–5
cleft, narrow segments, tip segment
longer than laterals; alternate.
Habitat: moist meadows,
roadsides. FL: May–June.

EASTERN FIGWORT
Scrophularia marilandica
FIGWORT FAMILY   SCROPHULARIACEAE

Perennial, 7' tall, bearing large branched cluster of irregular maroon-green flowers at top of leafy smooth stem. Flower $\frac{1}{4}$" long, corolla tube 2-lipped, upper long lip with 2 small lobes projects forward as hood; lower short lip 3-lobed with middle lobe bent down. Calyx small, toothed. Stamens 4 fertile, 1 sterile, on roof of upper lip. Ovary superior, short style extends above throat. Fruit cone-shaped capsule. Stem 4-angled, grooved, glandular. Leaves 6" long, ovate, pointed, toothed, petioled; opposite. Habitat: edge of moist woods. FL: July–September.

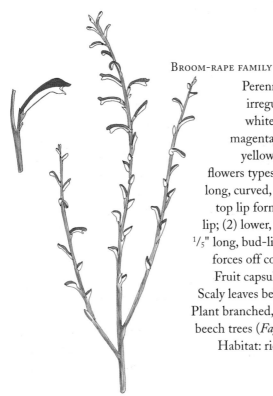

BEECHDROPS
Epifagus virginiana
BROOM-RAPE FAMILY   OROBANCHACEAE

Perennial, 18" tall, bearing irregular tubular maroon-white flowers along entire magenta-blotched stem with yellowish scale leaves. Two flowers types: (1) upper, male $\frac{1}{2}$" long, curved, 2-lipped, 4 stamens; top lip forms hood over bottom lip; (2) lower, small female flowers $\frac{1}{5}$" long, bud-like, developing ovary forces off corolla for pollination. Fruit capsule, many small seeds. Scaly leaves below flowers on stem. Plant branched, parasitic on roots of beech trees (*Fagus*), hence its name. Habitat: rich beech woods. FL: August–September.

CARDINAL-FLOWER
Lobelia cardinalis

BELLFLOWER FAMILY   CAMPANULACEAE

Perennial, 5' tall, bearing cluster of large irregular red flowers at top of leafy smooth stem with milky sap. Flower 2" long, corolla tube split half length, 2-lipped, upper lip with 2 erect narrow lobes and lower lip with 3 wider spreading lobes. Calyx short, 5 long teeth. Stamens 5, encircle style. Ovary inferior, style projects above throat. Fruit capsule. Flower on short stalk subtended by bract. Leaves 4" long, lanceolate-oblong, toothed, pointed, short petiole or sessile, smooth; alternate. Habitat: wet soil, creeksides. FL: July–September.

WILD HONEYSUCKLE
Lonicera dioica

HONEYSUCKLE FAMILY   CAPRIFOLIACEAE

Perennial vine, 10' long, bearing whorls of irregular tubular bright scarlet flowers at top of stem above perfoliate leaves. Flower $3/4$" wide, corolla hairy inside, deeply 5-lobed, lower 2 only shallow-lobed, tube base swollen on one side. Stamens 5, red filaments, yellow anthers, extend above corolla. Ovary inferior, style above corolla. Fruit berry. Stem smooth, wraps around host for support. Leaves: upper pair perfoliate under flowers; others, 5" long, elongate-oval, smooth, entire, sessile; opposite. Habitat: moist woods. FL: May–June.

ORANGE HAWKWEED
Hieracium aurantiacum

DAISY FAMILY   ASTERACEAE

Perennial, 18" tall, bearing red-orange flower-heads at top of sticky leafless hairy stalk above basal leaves. Head $^3/_4$" wide, ray flowers only. Corolla square-tipped, 5-notched. Calyx pappus of bristles. Stamens 5, encircle style. Ovary inferior, 1 pistil, lobed style. Fruit cylindrical achene tipped with pappus. Involucral bracts with black glands. Head stalk glandular, scale leaf at mid-height, exudes milky sap. Leaves basal 5" long, lanceolate, hairy. Habitat: fields, roadsides; introduced. FL: June–August. Yellow H., *H. caespitosum,* similar but with yellow flowers. Habitat: fields, roadsides; introduced.

DAY-LILY
Hemerocallis fulva

LILY FAMILY   LILIACEAE

Perennial, 4' tall, bearing several large funnel-shaped orange flowers at top of smooth leafless stalk. Flower $3^1/_2$" wide, 6 erect spreading tepals often with maroon patch near yellow throat, 3 inner with wavy margins and central rib. Stamens 6, brown anthers, long curved filaments. Ovary superior, 3 chambers, long slender style with small globose stigma. Fruit pod; seeds rare. Flower open only 1 day. Leaves basal, linear, keeled, arranged as fan; shorter than floral stalk. Habitat: homesteads, roadsides; introduced. FL: July–August.

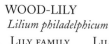

WOOD-LILY
Lilium philadelphicum
LILY FAMILY   LILIACEAE

Perennial, 3' tall, bearing 1–several large erect orange-red flowers at top of leafy smooth stem with narrow whorled leaves. Flower 4" wide, perianth 6 separate tepals, narrow to clawed base, spotted with black dots. Stamens 6, pale-orange, black anthers. Ovary superior, 1 pistil, long stigma-style, black spots on style. Fruit long capsule. Leaves 4" long, narrow-ovate; 4–7 at node. Habitat: moist-dry sandy thickets. FL: May–June.

SPOTTED CORAL-ROOT
Corallorhiza maculata
ORCHID FAMILY   ORCHIDACEAE

Perennial, 16" tall, bearing several irregular maroon flowers with white lip on leafless reddish stem. Flower ³⁄₈" long, petals 3, spotted purple, the lower lip tongue-like, 3-lobed with middle lobe largest; lateral petals curve in over lip. Maroon sepals 3, oblong, the upper sepal forms hood over lip, lateral sepals spreading. Stamen anthers 2, fused with style to form column above stigma; anther lid covers anther sacs. Ovary inferior, yellow, nodding. Fruit capsule. Stem base dull red, middle pink, top yellow. Red scale leaves, 1–2, sheath stem. Habitat: rich woods. FL: June–August.

RAM'S HEAD LADY-SLIPPER
Cypripedium arietinum

Perennial, 10" tall, bearing irregular downy maroon-pink flower with inflated conical pouch on leafy hairy stem. Flower $^3/_8$" long, petals 3, lower lip inflated conical pouch with white rim, veined dark-pink; lateral 2 petals narrow, project forward, maroon veins. Sepals 3, upper broad, erect, arching, maroon-striped; lateral 2 narrow, maroon-tipped. Stamen anthers 2, fused with style to form column that projects into lip; staminode covers stigma. Ovary inferior. Fruit capsule. Long bract below flower. Leaves 4" long, lanceolate, clasping. Habitat: pine woods. *Threatened.* FL: May–June.

Brown

FLOWERS

DWARF MISTLETOE
Arceuthobium pusillum

MISTLETOE FAMILY   VISCACEAE

Perennial, ³/₄" tall, parasitic, bearing small
brown flowers in axils of scale leaves and
inducing witches' brooms on black spruce trees.
Male and female plants often on different trees.
Female flower (right): perianth of tepals, 2-lobed,
ovary inferior with short stigma-style, fruit a
berry. Male flower (left): perianth of
spreading tepals, 3-lobed, each
lobe with a sessile yellow
stamen-anther. Stem short,
yellow-brown, sometimes
branched. Leaves are scales,
opposite, golden brown.
Habitat: branches of black
spruce. FL: June–July.

SQUAW-ROOT
Conopholis americana

BROOM-RAPE FAMILY   OROBANCHACEAE

Perennial, 8" tall, bearing crowded
spike of irregular tubular yellow-
brown flowers on fleshy yellow
stem with brown scale leaves.
Flower ¹/₂" long, corolla 2-
lipped, upper lip forming hood
over 3-lobed lower lip. Calyx
tube 5-toothed, ³/₄ corolla
length. Stamens 4, protrude
from corolla. Ovary superior, 1
pistil, style length of corolla.
Fruit capsule. Brown bract below
flower. Scale leaves alternate.
Plant parasitic on oak or hemlock
roots. Habitat: medium hemlock,
oak woods. FL: May–June.

SKUNK-CABBAGE
Symplocarpus foetidus
ARUM FAMILY   ARACEAE

Perennial, 1' tall, bearing mottled brown spathe around yellow-green malodorous flower spadix on leafless stalk with adjacent leaves. Flowers many, small, cover entire spadix. Flowers $1/4$" wide, perfect of 4 erect tepals. Stamens 4. Ovary embedded in spadix, thick style. Fruit seed covered by tepals, mature embedded in spadix, multiple fruit. Leaves basal 2' long, petioled, oval, heart-shaped, smooth, entire; enlarge after flowering. Habitat: swamps. FL: March–April.

NARROW-LEAVED CAT-TAIL
Typha angustifolia
CAT-TAIL FAMILY   TYPHACEAE

Perennial, 7' tall, bearing small brown male and female flowers crowded on cylindrical spike at top of floral stalk. Flower spike 8" long, male and female regions separated by gap. Male: 2–5 stamens fused along filaments, sepals and petals modified into bristles. Female: 1 pistil sessile to stem, linear stigma, sepals and petals modified into bristles. Males drop off after shedding pollen leaving tip of stalk bare. Fruit achene subtended with white hairs. Leaves $3/4$" wide, grass-like, ensheath stalk base; opposite. Habitat: marshes. FL: May–July

COMMON CAT-TAIL
Typha latifolia

<small>CAT-TAIL FAMILY   TYPHACEAE</small>

Perennial, 8' tall, bearing small brown male and female flowers crowded into cylindrical spike at top of floral stalk. Male above female flowers, many along 10" spike. Male: 2–5 stamens fused along filaments, sepals and petals modified into bristles. Female: 1 pistil nearly sessile to stalk, spatulate stigma, sepals and petals modified into bristles.

Males drop off after shedding pollen leaving tip of stalk bare. Fruit achene with white hairs. Leaves 1" wide, sword-like, pointed, ensheath stalk base; opposite. Habitat: marshes with flowing water. FL: May–July.

White

FLOWERS

FRAGRANT WATER-LILY
Nymphaea odorata
WATER-LILY FAMILY   NYMPHAEACEAE

Perennial, aquatic, 4' tall, bearing large fragrant floating white flower at top of long stalk with adjacent leaves. Flower 5" wide, petals many, decreasing in size toward many yellow stamens. Sepals 4, green. Ovary superior, many pistils united, their stigmas radiating from the top. Fruit large fleshy berry; ripens under water. Flower opens early morning, closes by noon. Underwater rhizome forms leaves and flowers. Leaves 10" round, cleft from base to petiole; green above, purplish below. Habitat: quiet water, ponds. FL: June–September.

WHITE BANEBERRY
Actaea alba
BUTTERCUP FAMILY   RANUNCULACEAE

Perennial, 2' tall, bearing dense long-stalked cluster of small white flowers on stem with much-divided leaf. Flower 1/4" wide, petals 4–10, spreading, lanceolate. Sepals 3–5, fall off as flower opens. Stamens many. Ovary superior, 1 pistil, broad sessile 2-lobed stigma. Fruit cluster waxy white berries with black spot, on thick red stalk; poisonous. Leaf, single large 2–3× divided, leaflets 3" long, ovate, pointed, toothed, smooth underside, terminal leaflet stalked. Habitat: rich woods. FL: May–June. Red B., *A. rubra,* similar but fruit red on slender stalk, leaflet underside downy. Habitat: rich rocky woods.

CANADIAN ANEMONE
Anemone canadensis

BUTTERCUP FAMILY   RANUNCULACEAE

Perennial, 2' tall, bearing large somewhat irregular white flower on long stalk at top of leafy hairy stem. Flower $1^1/_2$" wide, petals none. Sepals 5, white, broad, sometimes unequal. Stamens many, yellow, short. Each ovary superior, many pistils, long styles. Fruit flat, minutely hairy achenes in woolly globose head. Basal leaves 4" wide, hairy, long-petioled, deeply palmately cleft, toothed. Upper stem leaves, sessile, deeply 3-parted, toothed; single pair or whorl of 3. Habitat: sandy shores, wet meadows. FL: May–August.

THIMBLEWEED
Anemone virginiana

BUTTERCUP FAMILY   RANUNCULACEAE

Perennial, 3' tall, bearing single large showy white flower at top of long stalk on leafy hairy stem. Flower $1^1/_4$" wide, petals none. Sepals 5 (9) white, pointed. Stamens many, yellow. Each ovary superior, many pistils. Fruiting head oval $3/_4$" wide, thimble-like, achenes woolly, numerous. Leaves 3 in whorl midway up stem. Leaves petioled, deeply cleft into 3 lobes each often again lobed, toothed, taper to base. Center leaflet has curved outer edges. Habitat: rich rocky woods, streambanks. FL: June–July. *A. riparia*, center leaflet has straight wedged-shaped edges, fruit head narrow.

WOOD-ANEMONE
Anemone quinquefolia

BUTTERCUP FAMILY   RANUNCULACEAE

Perennial, 8" tall, bearing single large white flower at top of stalk above whorl of 3 deeply cleft leaves. Flower 1" wide, petals none. Sepals 4–9, white, oval, often pink underside. Stamens many, white. Ovary superior, several pistils and styles. Fruit minutely hairy achenes, in cluster. Stem smooth, often reddish, whorl of three petioled leaves below long flower stalk; often, solitary basal leaf present. Leaves palmately divided into 3–5 sharply toothed, narrow 1¼" long segments. Habitat: borders of rich woods, open woods. FL: April–June.

RUE-ANEMONE
Anemonella thalictroides

BUTTERCUP FAMILY   RANUNCULACEAE

Perennial, 8" tall, bearing several stalked white flowers elevated above whorl of 2–3 compound leaves with 3 blunt-lobed leaflets. Flower 1" wide, petals none. White sepals 5 or more, petal-like. Stamens many, orange-yellow. Ovary superior, several pistils, short round stigma. Fruit globose head of several ribbed spindle-shaped achenes, persistent stigma. Stem slender, reddish, smooth. Basal leaf long-petioled, twice three-parted, leaflet round, three blunt lobes at top. Stem leaves whorled, 3–4 blunt lobes at top. Habitat: rich moist or dry woods. FL: April–June.

VIRGIN'S BOWER
Clematis virginiana

Perennial vine, 10' long, bearing small clusters of large white flowers in axils on leafy stem. Separate plants with 1" wide male or female flowers; petals none. Male: sepals 4–5 white petal-like, many long stamens. Female: sepals 4–5 white, petal-like, several staminodes, many superior pistils with long style, fruit achenes in globose head, each achene with 1" persistent plumose style. Stem and petioles encircle host for support. Leaves petioled, compound with 3 sharply toothed oval petioled leaflets 2" long; opposite. Habitat: moist thickets. FL: July–September.

GOLDTHREAD
Coptis trifolia
BUTTERCUP FAMILY   RANUNCULACEAE

Perennial, 6" tall, bearing solitary white flower on long slender basal stalk with adjacent basal 3-foliate glossy leaves. Flower $5/8$" wide, petals none. Sepals 4–7, white, petal-like. Stamens many, yellow, some possess basal nectary. Ovary superior, several pistils each on short stalk, short style. Fruit follicle, persistent style, shiny black seeds. Stem creeping yellow rhizome. Leaves from rhizome, shiny, 3-lobed on long petiole; leaflets oval, $5/8$" wide, toothed, smooth, round tip. Habitat: wet mossy woods, bogs. FL: May–July.

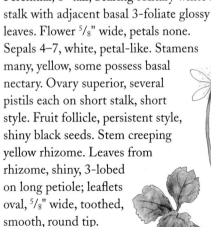

SHARP-LOBED HEPATICA
Hepatica acutiloba

BUTTERCUP FAMILY   RANUNCULACEAE

Perennial, 6" tall, bearing solitary white flower on basal hairy leafless stalk with adjacent 3-lobed leaves. Flower 1" wide, petals none. Sepals 5–12, white-pink, $^3/_4$" wide; pointed bracts below sepals. Stamens many, white. Ovary superior, several pistils, short styles. Fruit hairy achene; bracts persist under achenes. Leaves basal 3" wide, pointed, round, 3-lobed below middle of blade, hairy petiole, smooth blade. Habitat: rich upland woods. FL: March–June.

Round-lobed H., *H. americana*, similar but leaves round-tipped, not lobed to middle of blade. Habitat: moist upland woods.

WHITE WATER-CROWFOOT
Ranunculus trichophyllus

BUTTERCUP FAMILY   RANUNCULACEAE

Perennial aquatic, 2' long, bearing white flower at top of emerged stalk from submerged stem with feathery leaves. Flower $^5/_8$" wide, petals 5, spreading, basal nectary. Sepals 5, small. Stamens many, yellow. Each ovary superior, several pistils, short styles. Fruit globose head of achenes with minute stylar beak. Stem submerged, long, branched. Leaves $1^1/_2$" long, submerged, much dissected into fine filaments, petiole as long as stipule. Habitat: limestone pools along lakes; rare. FL: May–July.

MAY-APPLE
Podophyllum peltatum

Perennial, 20" tall, bearing single large pungent nodding white flower in crotch between two leaves. Flower 2" wide, petals 6–9, waxy. Sepals 6, fall when petals open. Stamens 6–9, yellow anthers. Ovary superior, 1 pistil, large sessile stigma. Fruit large acidic globose berry. Stem rhizome producing smooth flower stalk with 2 opposite leaves and flower in crotch, or vegetative stalk with only one leaf. Leaves 12" wide, round, umbrella-like, deeply palmately lobed. Forms dense stands in woods. Habitat: rich damp woods. FL: April–June.

MOONSEED
Menispermum canadense

Moonseed family   Menispermaceae

Perennial vine, 10' long, bearing small clusters of small whitish-green flowers in axils of leafy woody stem with peltate leaves. Male and female plants. Male flower: $^3/_{16}$" wide, petals 4, sepals 4 pointed and longer than petals, stamens many with yellow anthers (shown). Female flower: $^3/_{16}$" wide, petals 4, sepals 4, longer than petals, 3 superior ovaries, short style. Fruit blue drupe, crescent seed flat, poisonous. Stem wraps around host for support. Leaves 6" wide, round, petiole peltate near margin, 3–7 shallow lobes; alternate. Habitat: wetland wood thickets. FL: June–July.

BLOODROOT
Sanguinaria canadensis

POPPY FAMILY   PAPAVERACEAE

Perennial, 1' tall, bearing large white flower on smooth leafless stalk with large basal leaf that exudes colored sap. Flower 2" wide, petals 8, oblong. Sepals 2, fall when flower opens. Stamens many, yellow anthers. Ovary superior, 1 pistil, 2-lobed stigma-style. Fruit pointed elongated pod, many seeds. Leaves 8" long, round, palmately scalloped into 5–9 lobes, heart-shaped base, smooth, conspicuously veined. Stem rhizome sends up individual leaves and flower stalks, exudes yellow-orange sap. Habitat: rich woods. FL: April–May.

SQUIRREL-CORN
Dicentra canadensis

FUMITORY FAMILY   FUMARIACEAE

Perennial, 10" tall, bearing several fragrant nodding irregular white flowers on leafless stalk shorter than its leaves. Flower $5/8$" wide, petals 4, outer 2 together form parallel inflated heart-shaped spurs and extended petal lips, inner 2 are hooded and arched over stamens. Sepals 2, small, quickly shed. Stamens 6, yellow. Ovary superior, 1 style, 2-lobed stigma. Fruit oblong capsule, black ariled seeds. Stem small tuber. Leaves 6" long, divided into deeply cut leaflets, smooth, long petiole from tuber, gray-green above, exude yellow sap. Habitat: rich woods. FL: April–May.

DUTCHMAN'S BREECHES
Dicentra cucullaria

FUMITORY FAMILY   FUMARIACEAE

Perennial, 10" tall, bearing several fragrant nodding irregular white pantaloon-shaped flowers on leafless stalk above leaves. Flower $5/8$" wide, petals 4, outer 2 together form V-shaped inflated spurs with 2 flared yellow cup-like petal lips, inner 2 petals hooded and arched over stamens. Sepals 2, quickly shed. Stamens 6, yellow. Ovary superior, 1 style, 2-lobed stigma. Fruit oblong capsule, black ariled seeds. Stem small tuber. Leaves 6" long with much-dissected leaflets, smooth, long petiole from tuber, gray-green, exude yellow sap. Habitat: rich woods. FL: April–May.

WHITE CAMPION
Silene latifolia

PINK FAMILY   CARYOPHYLLACEAE

Annual or biennial, 3' tall, bearing fragrant large hairy flask-shaped white flowers on branched sticky leafy stem. Separate male and female plants. Male and female flowers: 1" wide, 5 deeply cleft petals, calyx sticky hairy bladder constricted to throat (upper). Male: 10 stamens, anthers at throat, narrow bladder with red-streaked veins. Female: superior ovary, 5 styles extend above petals, much-inflated green bladder with pink stripes (lower). Fruit capsule in bladder. Leaves 3" long, ovate, entire; opposite. Habitat: disturbed areas; introduced. FL: July–October.

BLADDER CAMPION
Silene vulgaris

PINK FAMILY   CARYOPHYLLACEAE

Perennial, 2' tall, bearing large white flask-shaped flowers on branches near top of smooth leafy stem. Male or female plants. Both male and female flower 1" wide, 5 deeply cleft petals, calyx an inflated smooth maroon bladder with faint netted venation, scattered hairs. Male: 10 short stamens. Female: superior ovary, 2–3 styles extend above petals. Fruit capsule with inflated calyx, seeds warty. Upper stem smooth, lower slightly sticky. Leaves 3" long, narrow-ovate, smooth, entire, pointed, sessile to clasping; opposite. Habitat: dry disturbed areas; introduced. FL: May–August.

COMMON CHICKWEED
Stellaria media

PINK FAMILY   CARYOPHYLLACEAE

Weak branched annual, 6" tall, bearing clusters of small white flowers with deeply lobed petals at top of leafy stem. Flower $1/4$" wide, petals 5, deeply cleft appearing like 10. Sepals 5, longer than petals, hairy. Stamens 5, black anthers. Ovary superior, 3 styles. Fruit capsule, many seeds. Stem much branched forming mats, 1 or several rows of hairs along stem. Opposite leaves 1" long. Lower leaves ovate, petioled, smooth, entire; upper leaves lanceolate, smooth, entire, sessile. Habitat: disturbed areas, lawns; introduced. FL: May–November.

BLACK BINDWEED
Polygonum convolvulus
SMARTWEED FAMILY   POLYGONACEAE

Annual vine, 5' long, bearing
cluster of small white flowers
on long axillary stalks on leafy
reddish stem. Flower $^{1}/_{8}$"
wide, perianth 5 tepals fused
at base, rough surface.
Stamens 10, short. Ovary
superior, 1 short style. Fruit
diamond-like achene (below).
Perianth midvein ridged, equals
size of achene. Stalk unbranched, small leaf in lower flower cluster.
Leaves 2" long, arrowhead base, pointed, smooth, entire, stipule
smooth, petioled; alternate. Habitat: moist woods; introduced. FL:
June–September.
Fringed B., *P. cilinode,* similar but stalk branched, stipule fringed at
node.
False Buckwheat, *P. scandens,* similar but tepals smooth, no leaf on
flower stalk. Habitat: openings in moist woods.

JAPANESE KNOTWEED
Polygonum cuspidatum
SMARTWEED FAMILY   POLYGONACEAE

Perennial, 6' tall, bearing large wide clusters of
small white flowers at top of woody hollow stem
with large ovate leaves. Separate male and
female plants. Male flower $^{1}/_{8}$" wide, perianth 5
tepals, 8 stamens, aborted ovary with 3 styles.
Female flower $^{1}/_{8}$" wide, perianth 5 tepals,
aborted stamens, 3-lobed round superior
ovary, 3 styles. Fruit 3-lobed black achene,
3 wings formed from inner tepals. Stem
smooth, woody, nodes solid. Leaves 6"
long, ovate, entire, smooth, petioled;
alternate. Habitat: roadsides,
disturbed areas; introduced.
FL: August–September.

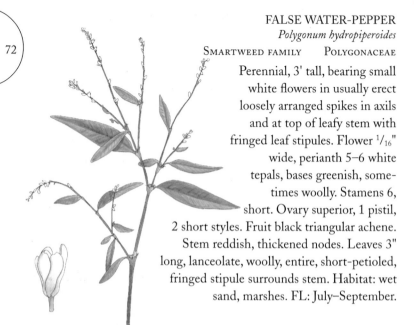

FALSE WATER-PEPPER
Polygonum hydropiperoides
Smartweed family   Polygonaceae

Perennial, 3' tall, bearing small white flowers in usually erect loosely arranged spikes in axils and at top of leafy stem with fringed leaf stipules. Flower $^1/_{16}$" wide, perianth 5–6 white tepals, bases greenish, sometimes woolly. Stamens 6, short. Ovary superior, 1 pistil, 2 short styles. Fruit black triangular achene. Stem reddish, thickened nodes. Leaves 3" long, lanceolate, woolly, entire, short-petioled, fringed stipule surrounds stem. Habitat: wet sand, marshes. FL: July–September.

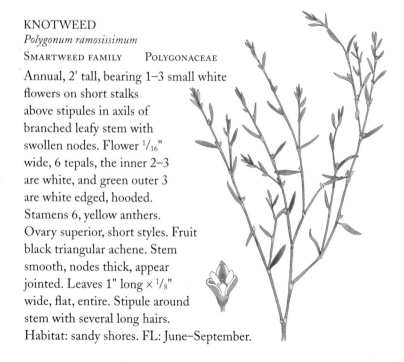

KNOTWEED
Polygonum ramosissimum
Smartweed family   Polygonaceae

Annual, 2' tall, bearing 1–3 small white flowers on short stalks above stipules in axils of branched leafy stem with swollen nodes. Flower $^1/_{16}$" wide, 6 tepals, the inner 2–3 are white, and green outer 3 are white edged, hooded. Stamens 6, yellow anthers. Ovary superior, short styles. Fruit black triangular achene. Stem smooth, nodes thick, appear jointed. Leaves 1" long × $^1/_8$" wide, flat, entire. Stipule around stem with several long hairs. Habitat: sandy shores. FL: June–September.

EUROPEAN FIELD-PANSY
Viola arvensis

VIOLET FAMILY   VIOLACEAE

Perennial, 1' tall, bearing nodding irregular white flower on long stalk in axils of leafy hairy stem. Flower ¹/₄" wide, petals 5, broad, upper 2 with purple tinged tip, lateral 2 spread wide, lower petal with several purple veins at yellow throat. Sepals 5, lanceolate, equal or longer than petals. Stamens 5. Ovary superior, 1 pistil, style with globose tip. Fruit elliptic capsule, brown seeds (right). Stem angular, reflexed hairs on angles. Leaves 1¹/₂" long, lanceolate, petioled, broadly toothed; stipule lyre-shaped, leaf-like. Habitat: disturbed areas, roadsides; introduced. FL: May–September.

SWEET WHITE VIOLET
Viola blanda

VIOLET FAMILY   VIOLACEAE

Perennial, 5" tall, bearing fragrant nodding irregular white flowers on reddish stalks separate from leaves. Flower ¹/₂" wide, petals 5, upper 2 narrow, somewhat reflexed, twisted, purple-veined at base; lower 3 petals with hairs at throat. Sepals 5, lanceolate, small. Stamens 5. Ovary superior, 1 pistil, 1 style. Fruit ¹/₄" oval purple capsule, brown seeds. Rhizome forms separate flower stalks and leaves. Leaf blade 2" wide, heart-shaped, not 2× as long as wide, pointed, shallow-toothed, dark-green, satiny sheen. Habitat: moist shaded slopes, woods. FL: April–May.

TALL WHITE VIOLET
Viola canadensis

VIOLET FAMILY   VIOLACEAE

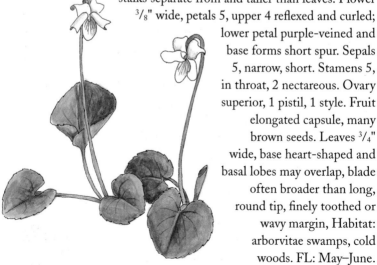

Perennial, 1' tall, bearing fragrant irregular white flowers on stalk in axils on leafy smooth stem. Flower not higher than leaves. Flower 1" wide, petals 5, inside white, outside violet, yellow throat. Petals spreading, lower lip wide and spurred; purple veins near base on laterals and lower petals. Sepals 5, small. Stamens 5, 2 produce nectar for spur. Ovary superior, short style. Fruit capsule, brown seeds. Leaf blades 2" wide, petioled, weakly heart-shaped base, toothed, pointed; stipules small and entire. Habitat: moist woods. FL: April–June.

KIDNEY-LEAVED VIOLET
Viola renifolia

VIOLET FAMILY   VIOLACEAE

Perennial, 4" tall, bearing nodding irregular white flowers on hairy stalks separate from and taller than leaves. Flower $^3/_8$" wide, petals 5, upper 4 reflexed and curled; lower petal purple-veined and base forms short spur. Sepals 5, narrow, short. Stamens 5, in throat, 2 nectareous. Ovary superior, 1 pistil, 1 style. Fruit elongated capsule, many brown seeds. Leaves $^3/_4$" wide, base heart-shaped and basal lobes may overlap, blade often broader than long, round tip, finely toothed or wavy margin, Habitat: arborvitae swamps, cold woods. FL: May–June.

GARLIC-MUSTARD
Alliaria petiolata
MUSTARD FAMILY   BRASSICACEAE

Biennial, 3' tall, bearing dense terminal clusters of white flowers at top and in axils of leafy smooth stem with garlic odor. Flower $^1/_3$" wide, petals 4, spatulate, spreading, narrow to claw. Sepals 4, small. Stamens 6, two short, 4 long, gland at filament base. Ovary superior, 1 pistil, short style. Fruit 2" long, erect 4-angled pod, elongates with petals at base, many seeds. Leaves 3" long, triangular, coarse teeth, petioled; alternate. Habitat: roadsides, fields, disturbed areas; introduced. Very weedy; excellent pot-herb. FL: May–June.

HAIRY ROCK-CRESS
Arabis hirsuta
MUSTARD FAMILY   BRASSICACEAE

Biennial, 30" tall, bearing terminal cluster of white flowers on leafy, finely hairy stem with basal leaves. Flower $^3/_{16}$" long, petals 4, spreading. Sepals 4, half length of petals. Stamens 6, two shorter than others. Ovary superior, 1 pistil, short style. Fruit 2" long, flat pod, erect, long stalk, winged seeds in 1 row. Basal leaves $^3/_4$" long, elliptical, branched hairs, toothed, blunt tips. Stem leaves smaller, clasp stem, widely spaced, toothed, sessile; alternate. Habitat: rocky disturbed soil. FL: May–July.

Tower-mustard, *A. glabra*, similar but stem leaves smooth, cylindrical pod erect, seeds in 2 rows. Habitat: dry soil, fencerows.

LYRE-LEAVED ROCK-CRESS
Arabis lyrata

MUSTARD FAMILY   BRASSICACEAE

Biennial, 16" tall, bearing
terminal cluster of white
flowers on leafy stem with
basal rosette of lobed leaves.
Flower $1/4$" wide, petals 4,
spatulate, spreading, narrow to
claw. Sepals 4, small. Stamens
6, two short, glands at base.
Ovary superior, 1 pistil, style
short. Fruit $1^1/4$" long, upcurved
thin pod, elongates with petals at
base; oval pitted seeds. Lower stem
hairy, some branched hairs, leaves of
rosette $1^1/4$" long, pinnately lobed, hairy.
Leaves on stem narrow, spatulate, taper to base, sessile,
entire; alternate. Habitat: sandy areas, cliffs, ledges. FL: April–May.

HOARY ALYSSUM
Berteroa incana

MUSTARD FAMILY   BRASSICACEAE

Annual, 30" tall, bearing terminal cluster of small white flowers at top
and in axils of leafy stem. Flower $1/4$" wide,
petals 4, deeply cleft, spreading, narrow to
basal claw. Sepals 4, small. Stamens 6, two
short, 4 long. Ovary
superior, 1 pistil, long
stigma-style. Fruit short-
stalked, erect, elliptical,
finely hairy beaked pod, flat
brown seeds. Stem, leaves covered
with short, erect, branched hairs;
velvety. Leaves 2" long, lanceolate,
smooth, entire, sessile; alternate. Habitat: dry soil,
roadsides; introduced. FL:
June–September.

SHEPHERD'S PURSE
Capsella bursa-pastoris
MUSTARD FAMILY   BRASSICACEAE

Annual, 2' tall, bearing long terminal clusters of small white flowers on leafy stem and forming heart-shaped fruits. Flower $^1/_{16}$" long, petals 4, spatulate, forming cross. Sepals 4, oblong, white-edged, shorter than petals. Stamens 6, minute glands at base. Ovary superior, 1 pistil, short style. Fruit flat, heart-shaped, round margins, tip notched with persistent style, seed triangular. Basal leaves 4" long, rosette, pinnately toothed, hairy, dandelion-like. Stem leaves small, arrow-shaped, clasp somewhat hairy stem; alternate. Habitat: disturbed areas; introduced. FL: May–October.

BROAD-LEAVED TOOTHWORT
Cardamine diphylla

MUSTARD FAMILY   BRASSICACEAE

Perennial, 1' tall, bearing loose terminal cluster of large white flowers on stalk above pair of trifoliate leaves. Flower $^3/_4$" wide, petals 4, spatulate, narrow to claw, aging to pink. Sepals 4, linear, half length of petals. Stamens 6, two short with U-shaped basal gland. Ovary superior, 1 pistil, slender style. Fruit slender round 1" long beaked pod, wrinkled brown seeds. Upright stem smooth, bears 2 nearly opposite petioled trifoliate leaves below inflorescence, each oval leaflet pointed, coarse rounded teeth. Habitat: rich moist woods. FL: April–June.

PENNSYLVANIA BITTER-CRESS
Cardamine pensylvanica

MUSTARD FAMILY   BRASSICACEAE

Annual or biennial plant, 2'
tall, bearing terminal
cluster of small white
flowers on leafy stem with
pinnate leaves. Flower $1/8$"
wide, petals 4, upright.
Sepals 4, small. Stamens 6,
two short with U-shaped
basal gland. Ovary
superior, 1 pistil, slender
style. Fruit linear, upright, 4-
angled, it elongates with petals at
base (right). Basal leaves pinnately compound, tip leaflet round or 3-
lobed, smooth, short-stalked. Stem leaves petioled, with 2–15 elongated smooth leaflets, narrow to base; alternate. Habitat: swamps,
ponds, wet woods. FL: May–June.

DAME'S ROCKET
Hesperis matronalis

MUSTARD FAMILY   BRASSICACEAE

Perennial, 3' tall, bearing cluster of
large fragrant white flowers at top of
leafy hairy stem. Flower 1" wide,
petals 4, oval, spreading, narrow
to claw. Sepals 4, shorter than
claw, pink to green. Stamens 6,
two short, U-shaped glands at
base. Ovary superior, 1 pistil,
short lobed style. Fruit long pod
widely spreading on hairy stalk,
elongates with petals at base, angular
seeds. Stem, leaves with straight and
branched hairs. Leaves 4" long, ovate,
broadened lobed base, sessile or short-
petioled, toothed; alternate. Habitat: wood
edges; introduced. FL: May–July.

PEPPER-GRASS
Lepidium densiflorum
MUSTARD FAMILY   BRASSICACEAE

Annual, 2' tall, bearing long clusters of tiny white flowers in upper axils of leafy finely hairy stem with round flat fruits. Flower $1/16$" wide, petals 4, shorter than sepals. Sepals 4, white-edged. Stamens 2, project above sepals (above). Ovary superior, short style. Fruit dry flat pod, stylar beak in notch, 2 wrinkled winged seeds. Basal leaves 2" long, toothed, often dried or absent when in flower. Stem leaves 1" long, lanceolate, mostly entire; alternate. Habitat: roadsides, disturbed areas; introduced. FL: May–August.

WATER-CRESS
Rorippa nasturtium–aquaticum
MUSTARD FAMILY   BRASSICACEAE

Perennial aquatic, 1' tall, bearing clusters of small white flowers in axils of smooth stem with pinnate leaves. Flower $1/8$" wide, petals 4, spatulate, clawed. Sepals 4, maroon-tipped, half length of petals. Stamens 6, yellow, two short with basal glands. Ovary superior, short style, lobed stigma. Fruit slender upward curved 1" pod, black seeds. Stem slightly ribbed, floating or submerged, roots at nodes. Leaves 6" long, with 5–9 oval leaflets, finely toothed or wavy, terminal leaflet largest; alternate. Pungent odor. Habitat: flowing water, springs; introduced. FL: April–October.

FIELD PENNY-CRESS
Thlaspi arvense
MUSTARD FAMILY   BRASSICACEAE

Annual, 20" tall, bearing elongated clusters of small white flowers on leafy smooth stem and forming round flat fruits. Flower $^1/_8$" wide, petals 4, spatulate, flat top. Sepals 4, yellow-green, white edges, shorter than petals. Stamens 6, short, yellow anthers, basal glands. Ovary superior, 1 pistil, white style. Fruit $^3/_8$" wide, marginal wings, tip deeply notched, 4 or more seeds. Basal leaves petioled, often dried or absent. Stem leaves lanceolate, toothed, sessile, arrow-shaped base with reflexed wings; alternate. Habitat: disturbed areas; introduced. FL: June–September.

BEARBERRY
Arctostaphylos uva-ursi
HEATH FAMILY   ERICACEAE

Trailing perennial, 1' tall, bearing small terminal cluster of nodding bell-shaped white to pink flowers in axils among leathery leaves. Flower $^1/_4$" long, corolla oval with 5 small lobes at mouth. Calyx 5, short. Stamens 10, short, pubescent, with 2 awns. Ovary superior, 5-chambered, globose, persistent stigma-style extends to flower lip. Fruit fleshy red drupe, 5 nutlets. Horizontal stems form mat, flowers and leaves on erect stems. Leaves $1^1/_4$" long, shiny, entire, thick, paddle-shaped; alternate. Habitat: sandy or rocky soil, roadsides. FL: May–June.

LEATHERLEAF
Chamaedaphne calyculata
Heath family   Ericaceae

Perennial, 4' tall, bearing nodding white bell-shaped flowers along lower side of leafy branches. Flower $1/4$" long, corolla tube with small 5-lobed mouth. Calyx 5-toothed, short, and subtended by 2 small bracts. Stamens 10. Ovary superior, style extending to corolla lip. Fruit globular capsule. Stem somewhat shrubby, scaly. Leaves 2" long, leathery, oblong, entire, short petiole, dotted with small gray scales especially on underside; alternate. Older leaves bronze-yellow underside. Habitat: sphagnum bogs, wet evergreen areas. FL: March–July.

TRAILING ARBUTUS
Epigaea repens
Heath family   Ericaceae

Perennial, prostrate, bearing clusters of fragrant white flowers on short stalk in axils on leafy hairy stems. Flower $1/2$" wide, corolla tube cleft half its length to form 5 spreading lobes with hairy tube throat. Sepals 5, short, pointed, subtended by green hairy bracts. Stamens 10, included in throat. Ovary superior, 5-chambered, 5-lobed stigma at tip of slender style. Fruit globose pulpy capsule, brown pitted seeds. Stem woody. Leaves 3" long, oval, leathery, evergreen, short hairy petiole; alternate. Habitat: sandy moist shady woods near bogs. FL: February–May.

WINTERGREEN
Gaultheria procumbens
HEATH FAMILY   ERICACEAE

Trailing perennial, 8"
tall, bearing a few
nodding bell-shaped white
flowers on short stalk near top of
leafy stem with evergreen leaves.
Flower $^3/_8$" long, corolla tube with 5
small spreading lobes. Calyx 5-
pointed, saucer-like, whitish, often
subtended by bracts. Stamens 4, in
tube. Ovary superior, long style
evident after tube falls. Fruit cream-red
berry. Horizontal stem, flowers and leaves on
erect stems, leaves at tip. Leaves $1^1/_4$" long, elliptic, toothed, pointed,
smooth, fine hairs on underside, pink petiole; alternate. Habitat: moist
acid woods. FL: July–August.

LABRADOR-TEA
Ledum groenlandicum
HEATH FAMILY   ERICACEAE

Evergreen perennial, 4' tall, bearing rounded terminal clusters of white
flowers on hairy stem. Flower $^1/_2$" wide, on slender stalk, petals 5,
spreading. Calyx 5-lobed, short. Stamens 5–
7, as long as petals, white anther. Ovary
superior, 5-celled, minutely hairy,
elongated style. Fruit $^1/_4$" oval beaked
capsule on recurved stalk.
Leaves 2" long, linear,
entire, short-petioled,
dark green upper
surface, leathery;
alternate. Blade slightly
rolled in at margin; lower
surface covered with woolly
orange hair; crushed blade
fragrant. Habitat: wet shores,
bogs. FL: May–June.

SPOTTED WINTERGREEN
Chimaphila maculata

SHINLEAF FAMILY   PYROLACEAE

Perennial, 8" tall, bearing a few nodding waxy white flowers at top of stalk on stem with whorled leaves lined with white. Flower $^3/_4$" wide, petals 5, round, spreading, sometimes tips incurved. Sepals 5, small, spreading. Stamens 10, short, yellow anthers. Ovary superior, style short but grows in length as fruit enlarges, globose stigma. Fruit brown capsule. Stem creeping with erect brownish flower stems. Leaves 2" long, midrib white, ovate, shiny, bluish-green, petioled; whorled. Habitat: rich sandy woods. FL: June–August.

ONE-FLOWERED SHINLEAF
Moneses uniflora

SHINLEAF FAMILY   PYROLACEAE

Perennial, 6" tall, bearing one fragrant delicate nodding creamy white flower on bent flower stalk above small cluster of basal leaves. Flower $^3/_4$" wide, petals 5, oval, spreading. Sepals 5, small. Stamens 10, nearly length of petals, yellow anthers. Ovary superior, 5-celled, style elongate, 5-lobed stigma. Fruit capsule. Plant smooth, scale leaf on stalk. Basal leaves 1" long, round, finely toothed, thin; petiole length of blade. Habitat: damp woods, bogs. FL: July–August.

ONE-SIDED SHINLEAF
Pyrola secunda

SHINLEAF FAMILY   PYROLACEAE

Perennial, 8" tall, bearing several waxy fragrant somewhat nodding white flowers along only one side of bent stalk with scale leaf at midstalk. Flower $1/16$" wide, petals 5, with basal swellings, oblong and erect. Sepals 5, short, ovate. Superior ovary, 5-celled, subtended by 10-lobed disk; straight stigma-style longer than petals. Fruit capsule. Basal leaves $1/2$" wide, oval, finely toothed, shiny, distinctive venation, long-petioled. Habitat: moist wood, mossy bogs. FL: June–July.

PINESAP
Monotropa hypopithys

INDIAN PIPE FAMILY   MONOTROPACEAE

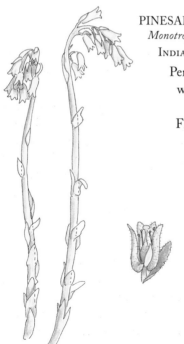

Perennial, 1' tall, bearing several fleshy waxy nodding white flowers at top of fleshy stem with scaly white bracts. Flower $1/2$" long, petals 4–5, woolly, 1 smaller than others. Sepals none. Stamens 5, short. Ovary superior, woolly, 1 oval pistil, short thick style, enlarged stigma. Fruit capsule, thread-like seeds. Flower subtended by hairy, toothed bract. Flowers erect after fertilization. Stem cream; blackens with age. Leaves $3/8$" long, translucent scales, smooth, entire; alternate. Mycotrophic. Habitat: moist, humus woods. FL: June–August.

INDIAN PIPE
Monotropa uniflora
INDIAN PIPE FAMILY   MONOTROPACEAE

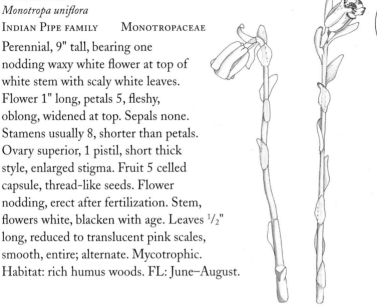

Perennial, 9" tall, bearing one
nodding waxy white flower at top of
white stem with scaly white leaves.
Flower 1" long, petals 5, fleshy,
oblong, widened at top. Sepals none.
Stamens usually 8, shorter than petals.
Ovary superior, 1 pistil, short thick
style, enlarged stigma. Fruit 5 celled
capsule, thread-like seeds. Flower
nodding, erect after fertilization. Stem,
flowers white, blacken with age. Leaves $1/2$"
long, reduced to translucent pink scales,
smooth, entire; alternate. Mycotrophic.
Habitat: rich humus woods. FL: June–August.

STARFLOWER
Trientalis borealis
PRIMROSE FAMILY   PRIMULACEAE

Perennial, 10" tall, bearing typically
two white flowers on slender
stalks arising
from whorl of
leaves. Flower
$1/2$" wide, corolla 7
pointed lobes. Calyx
7, small lobes. Stamens
7, yellow anthers
extend above corolla.
Ovary superior, 1
pistil, long style, globose
stigma. Fruit capsule, many black
reticulated seeds. Leaves 4" long,
lanceolate, taper at both ends, in whorl
of 5–9. Small scale leaf present
midway along stem. Habitat: cool wet
woodlands, bogs. FL: May–June.

DOGBERRY
Ribes cynosbati

GOOSEBERRY FAMILY   GROSSULARIACEAE

Perennial, 4' tall, bearing 1–3 white
flowers each above spiny ovary on stalk in
axils on leafy stem with scattered or no
spines. Flower ¼" wide, cup-shaped, 5 tiny
flat-topped petals. Sepals 5, on cup above
ovary, reflexed, longer than petals.
Stamens 5, shorter than sepals,
white filaments, green anthers.
Ovary inferior, glandular
spines, long-lobed style. Fruit
spiny red-purple berry. Leaf blade
1½" long, long petiole, 3–5 palmate
lobes, round teeth, finely hairy; alternate.
Habitat: rich moist woods. FL: May–June.
Northern Gooseb., *R. hirtellum,* similar but
long stamens equal sepal length, fruit spineless. Habitat: rocky woods.

TWO-LEAVED MITREWORT
Mitella diphylla

SAXIFRAGE FAMILY   SAXIFRAGACEAE

Perennial, 18" tall, bearing loose spike of
delicate white flowers on downy stalk above
pair of leaves. Flower ³⁄₁₆" wide, petals 5,
fringed, resembling snowflake. Sepals 5,
white, oval, pointed. Stamens 10, erect,
above petals, yellow. Ovary inferior, 1
pistil, 2 short styles. Fruit 2-beaked
capsule (mitre-like), black seeds. Stem
with basal petioled leaves, 3-
to 5-lobed, heart-shaped,
toothed, hairy. Flower stalk
possesses opposite pair of 3-
lobed sessile leaves at
midpoint. Habitat: rich wet
rocky woods, often growing
on moss. FL: April–June.

GRASS-OF-PARNASSUS
Parnassia glauca
Saxifrage family   Saxifrageaceae

Perennial, 20" tall, bearing one white green-
striped flower at top of smooth stalk above
basal leaf rosette. Flower 1" wide, petals 5,
oval, green veins. Sepals 5, small,
maroon. Stamens 5 fertile plus 3
staminodes each tipped with gland.
Ovary superior, 1 pistil, 4 stigmas. Fruit
4-parted capsule, winged seeds. Sessile leaf
near base of stalk. Basal leaves $2^1/_2$" wide,
blade longer than wide, heart-shaped,
entire, smooth, long-petioled. Habitat:
wet limy meadows. FL: July–October.

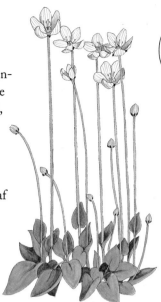

WHITE AVENS
Geum canadense
Rose family   Rosaceae

Perennial, 3' tall, bearing several white flowers each on long velvety
stalks in axils on hairy stem with 3-
foliate leaves. Flower $^5/_8$" wide,
petals 5, oval, spreading. Sepals 5,
recurved, pointed, as long as petals.
Stamens 10, yellow. Ovaries superior,
many on domed receptacle, long
jointed styles. Fruit many achenes,
many short bristles among achenes,
style breaks to form
hooked beak on
achene. Basal leaves
trifoliate, petioled,
elliptic leaflets 3" long,
toothed. Upper leaves
trifoliate or lobed,
elliptic, toothed,
petioled; alternate.
Habitat: moist woods.
FL: May–June.

WILD STRAWBERRY
Fragaria virginiana
ROSE FAMILY   ROSACEAE

Perennial, 6" tall, bearing several white
flowers on leafless stalk with basal
trifoliate leaves. Flower $3/4$" wide, petals
5, spreading. Sepals 5, small. Stamens
many, yellow. Ovary superior, many
pistils on domed receptacle. Fruit
globose, achene embedded in pit on
fleshy receptacle that enlarges into red
berry. Leaves taller than flower stalk,
long hairy petiole, leaflets $1^1/_2$" long,
toothed and terminal tooth shorter
than flanking teeth. Habitat:
roadsides, fields. FL: May–June.
Woodland S., *F. vesca*, similar but terminal tooth longest, fruit conical,
achene on surface of fruit. Habitat: woodland edges.

COMMON BLACKBERRY
Rubus allegheniensis
ROSE FAMILY   ROSACEAE

Perennial, 5' tall, bearing large clusters
of white flowers on spiny long arching
stem with trifoliate leaves. Flower $3/4$"
wide, petals 5. Sepals 5. Stamens
many, yellow. Ovary superior,
many pistils on receptacle. Fruit
black drupelets bearing achene,
receptacle separates with fruit.
Stem spines straight, sharp,
glandular. Leaflets 4" long, oval,
toothed, center leaflet long-
stalked; alternate. Habitat: dry
clearings. FL: May–July.
Red Raspberry, *R. idaeus*, stalk
with straight glandular bristles,
red fruit separates like cap. Black R., *R.
occidentalis*, black fruit separates like cap. Habitat: dry clearings.

THIMBLEBERRY
Rubus parviflorus

ROSE FAMILY   ROSACEAE

Perennial, 4' tall, bearing loose cluster of several large white flowers near top of hairy stem with palmately lobed leaves. Flower 1³/₄" wide, petals 5, broad, over ⁵/₈" long, spreading. Sepals 5, yellow glandular hairs, equal or longer than petals. Stamens many, yellow anthers. Ovary superior, many pistils on receptacle. Fruit ¹/₂" wide red edible cap of drupelets each bearing achene, separates from receptacle, persistent style. Stem woody, rough, hairy, glandular. Leaves 8" wide, maple-like, heart-shaped base, toothed, stipules; alternate. Habitat: open woods. FL: May–July.

DWARF RASPBERRY
Rubus pubescens

ROSE FAMILY   ROSACEAE

Perennial, 1' tall, bearing 1–3 small white flowers near top of nearly smooth stem with shiny 3- or 5-parted leaves. Flower ¹/₄" wide, petals 5, white, linear, twisted. Sepals 5, white recurved. Stamens many. Ovary superior, several pistils on receptacle. Fruit dark red drupelets each bearing achene, fruit adheres to receptacle. Leaves 3- or 5-foliate, long-petioled, leaflet 2" long, oval, sharply toothed above middle, oblong pointed stipules; alternate. Habitat: wet woods, bogs. FL: May–July.
Swamp Dewberry, *R. hispidus,* similar but wider spreading petals, slender reflexed prickles. Habitat: bogs.

NARROW-LEAVED MEADOWSWEET
Spiraea alba

ROSE FAMILY   ROSACEAE

Perennial, 5' tall, bearing elongated terminal
cluster of numerous fragrant white or slightly
pink flowers on leafy stem. Flower $\frac{1}{4}$" wide,
petals 5, spreading. Sepals 5, spreading,
small. Stamens many, long, filaments form
red central ring. Ovary superior, 5 pistils
and styles. Fruit dry follicle. Stem woody,
red at base, green above. Leaves 2"
long, lanceolate, widest above middle,
short-petioled, smooth, toothed;
alternate. Habitat: edge of moist
woods. FL: June–August.

WHITE PEA
Lathyrus ochroleucus

PEA FAMILY   FABACEAE

Perennial, 30" tall, bearing small axillary cluster of irregular white
flowers on short stalk in axils of smooth stem with compound leaves
and tendrils. Flower $\frac{5}{8}$" long, corolla 2-lipped, upper lip is broad and
notched, white laterals are large and incurved, lower
lip of 2 fused green lobes form keel. Calyx
lobes unequal, lower lobe
longer than upper.
Stamens 10. Ovary
superior, 1 pistil, hairy
style. Fruit smooth
pod. Leaves
pinnate, 3–5 pairs
of leaflets, oblong,
smooth, entire,
tendril at leaf tip,
large stipules; alternate.
Habitat: upland
woods, thickets;
introduced. FL:
May–July.

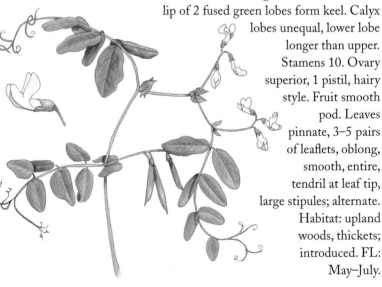

WHITE SWEET CLOVER
Melilotus alba

PEA FAMILY   FABACEAE

Annual, 6' tall, bearing long loose
spikes of small irregular white
flowers at top and in axils of smooth
branched stem with trifoliate leaves.
Flower $^3/_{16}$" long, petals clawed,
upper petal oblong, longer than 2
lateral and lower keel petals. Calyx
cup-shaped, green, 5 short
teeth. Stamens 10, yellow
anthers. Ovary superior, 1
short pistil. Fruit oval pod.
Leaves petioled, alternate;
leaflets linear, sessile, finely toothed, smooth,
small stipule. Habitat: fields; introduced. FL: June–September.
Yellow Sweet Clover, *M. officinalis*, similar with yellow flowers.
Habitat: fields.

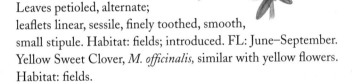

DWARF ENCHANTER'S NIGHTSHADE
Circaea alpina

EVENING-PRIMROSE FAMILY   ONAGRACEAE

Perennial, 10" inches, bearing widely
spaced small white flowers at top of leafy
stem. Flower $^1/_8$" wide, petals 2, deeply
lobed, fused into tube at their
base. Sepals 2, pink tips,
spoon-shaped, attached at
top of ovary. Stamens 2, long.
Ovary inferior, hairy, bilobed, 1
long style, lobed stigma. Fruit
dangles on long thin minutely
hairy stalk; capsule oval, not
marked with ridges, densely
covered with minute hooked hairs
that stick to fur, clothing. Leaves 2" long, broad,
often heart-shaped, toothed, pointed, petioled; opposite.
Habitat: moist woods, swamps, bogs. FL: June–August.

ENCHANTER'S NIGHTSHADE
Circaea lutetiana
EVENING-PRIMROSE FAMILY   ONAGRACEAE

Perennial, 2' tall, bearing widely spaced small white long-stalked flowers on arched spike at top of leafy stem. Flower $^3/_{16}$" wide, petals 2, deeply lobed, fused into tube at base. Sepals 2, partly reflexed, attached at top of ovary. Stamens 2. Ovary inferior, bilobed, hairy, 1 long style, globose stigma. Fruit dangles on long thin stalk, bilobed oval capsule marked with 10 ridges, covered with hooked hairs that stick to fur, clothing. Leaves 4" long, broad base, ovate, toothed, pointed, downy underside, petioled; opposite. Habitat: moist shady woods. FL: June–August.

NORTHERN WILLOW-HERB
Epilobium glandulosum
EVENING-PRIMROSE FAMILY   ONAGRACEAE

Perennial, 3' tall, bearing solitary white flower on an apparently long stalk in upper axils of sticky leafy stem. Flower $^1/_4$" wide, petals 4, notched, spreading. Sepals 4, pointed, $^1/_3$ length petals. Stamens 4, short. Ovary inferior, short style, globular stigma. Fruit 3" long, thin, twisted, capsule; seed tufted with brown silk. Stem downy, incurved hairs, somewhat branched above. Leaves 3" long, narrow-ovate, toothed, short-petioled, pointed, red margin; opposite. Habitat: bogs, wet soil. FL: July–September.

Marsh W-herb, *E. palustre*, similar habitat but smaller (1'), only 1 small flower, linear leaves. Habitat: marshes.

BUNCHBERRY
Cornus canadensis
<small>DOGWOOD FAMILY   CORNACEAE</small>

Perennial, 8" tall, bearing terminal cluster
of tiny flowers subtended by 4 white oval
bracts above whorl of leaves. White
bracts $3/4$" wide, broadly round,
pointed, below flowers. Flower
tiny, petals 4, yellowish. Sepals
4, minute. Stamens 4, small.
Ovary inferior, style with
globose stigma. Fruit $1/4$" red
drupe, several in cluster. Bracts
and flowers supported on long stalk above whorl
of 4–6 elliptic, pointed, 3" long leaves. Stem
unbranched, opposite pairs of scale leaves on
stem. Habitat: cool wet woods. FL: May–June.

BASTARD TOAD-FLAX
Comandra umbellata
<small>SANDALWOOD FAMILY   SANTALACEAE</small>

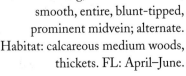

Perennial, 1' tall, bearing a
flat-topped terminal cluster
of small funnel-shaped white
flowers on leafy stem. Flower
$3/16$" wide, white tepals 5,
upright, pointed, forming ring on
top of ovary. Stamens 5, attached to
tepals with tuft of hair. Ovary
inferior, 1 glandular pistil, 1
style. Fruit juicy drupe. Stem
upright; partially parasitic on
roots of other plants. Leaves
$11/4$" long, narrow-lanceolate,
smooth, entire, blunt-tipped,
prominent midvein; alternate.
Habitat: calcareous medium woods,
thickets. FL: April–June.

NEW JERSEY TEA
Ceanothus americanus
BUCKTHORN FAMILY   RHAMNACEAE

Perennial, 3' tall, bearing small white flowers in oval clusters on long axillary stalks near top of leafy shrubby stem. Flower $1/8$" wide, petals 5, spreading, bucket-like lobe with a long claw. Sepals 5, triangular, between petal claws, infolded against ovary. Stamens 5, erect, red. Ovary superior, 3-celled, 3 styles. Fruit 3-lobed drupe. Leaves 3" long, ovate, smooth, toothed, sharply pointed; alternate. Nitrogen-fixing plant. Habitat: open woods, roadside clearings. FL: June–July.

SPIKENARD
Aralia racemosa
GINSENG FAMILY   ARALIACEAE

Perennial, 4' tall, bearing many small umbels of tiny white flowers on short stalk attached to stem with pinnate leaves. Flower $1/8$" wide, petals 5, reflexed. Sepals reduced. Stamens 5, white anthers make flower appear fuzzy. Ovary inferior, 5-celled, 1 pistil, style. Fruit purple drupe. Leaves 2× pinnately compound, leaflets heart-shaped, smooth, toothed, sharply pointed, petioled; appear opposite but arranged pinnately on 3 axes of leaf. Habitat: rich woods. FL: June–July.

DWARF GINSENG
Panax trifolius
<small>GINSENG FAMILY</small>   <small>ARALIACEAE</small>

Perennial, 8" tall, bearing delicate
solitary umbel of many tiny white
flowers on stalk above whorl of 3
palmately compound leaves.
Flower $^1/_{16}$" wide, petals 5,
tips curve up; basal nectary
a ring (shown). Sepals
none. Stamens 5, short.
Ovary inferior, 2–3
celled, long style. Fruit
yellow berry. Umbel
subtended by bract. Leaves
long-petioled, 3- to 5-palmately lobed;
leaflets 1" long, sessile, lanceolate, smooth,
finely toothed. Whorl of 3 leaves below stalk of umbel.
Habitat: moist rich woods. FL: April–June.

CARAWAY
Carum carvi

<small>CARROT FAMILY</small>   <small>APIACEAE</small>

Biennial, 3' tall, bearing compound
umbels of small white flowers at top and
in axils of hairy smooth stem with
pinnately dissected leaves. Flower $^1/_8$" wide,
corolla 5 lobes, tips infolded. Calyx
none. Stamens 5, longer than corolla
lobes. Ovary inferior, 2 styles flattened
laterally. Fruit oblong, ribbed, smooth
seed with short style. Membranous sheath
partially encloses leaf petiole and
flower stalk. Developing umbel
emerges from membranous sheath
during maturation. Umbels subtended
by linear bract. Leaflets $^1/_2$" long,
feather-like. Habitat: disturbed areas;
introduced. FL: June–August.

BULB-BEARING WATER-HEMLOCK
Cicuta bulbifera
CARROT FAMILY   APIACEAE

Perennial, 3' tall, bearing small umbels of small white flowers at top of leafy smooth stem with bulblets in leaf axils. Flower $1/16$" wide, petals 5. Sepals none. Stamens 5. Ovary inferior. Fruit, rarely formed; reproduces by bulblets. Small bract below umbel. Stem slender, branched, hollow, aromatic. Lower leaves 10" long, 2× pinnate, 5–7 leaflets, narrow $1/4$" wide or less, toothed, pointed. Upper stem leaves, strap-shaped with bulblets in axil; appear opposite. Poisonous if eaten. Habitat: swamps, marshes. FL: July–August.

HONEWORT
Cryptotaenia canadensis
CARROT FAMILY   APIACEAE

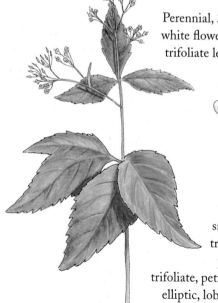

Perennial, 3' tall, bearing loose umbels of white flowers at top of smooth stem with trifoliate leaves. Umbellets few flowered. Flower $1/8$" wide, corolla 5-lobed, tips inturned. Sepals none. Stamens 5, extend above corolla. Ovary inferior, forked style short. Fruit flat, elongate, dark, smooth, short stylar tip. Umbel with small or no bract. Lower leaves trifoliate, 6" long, leaflets ovate, petioled, toothed. Stem leaves trifoliate, petioled or sessile, leaflet 2" long, elliptic, lobed, toothed, smooth. Habitat: mixed woods. FL: June–July.

WILD CARROT

Daucus carota

CARROT FAMILY APIACEAE

Biennial, 4' tall, bearing large umbels of scented white flowers on stalk on leafy hairy stem with large pinnately divided bract under umbel. Flower $1/8$" wide, petals 5, one larger than others. Sepals none. Stamens 5, yellow. Ovary inferior, short style. Fruit 2-seeded, flat, prickly hairs. Umbel compound, marginal flowers often larger than others; central umbel may have purple flower. Leaves 2' long, 2× pinnate, leaflets narrow segments. Umbel after pollination curls inward appearing like bird's nest. Origin of cultivated carrot. Habitat: disturbed areas; introduced. FL: May–October.

COW-PARSNIP

Heracleum lanatum

CARROT FAMILY APIACEAE

Perennial, 8' tall, bearing large compound umbels of white flowers on leafy hairy stem with large palmate leaves. Flower $1/4$" wide, corolla 5 lobes, irregular, deeply notched. Sepals none. Stamens 5, shorter than corolla. Ovary inferior, styles short. Fruit flat, elliptic, winged, smooth. Umbel long-stalked, subtending bract falls early leaving hairy ring. Umbellets with 15–30 flowers. Sheath partially encloses young leaves and flower stalk. Leaves trifoliate, leaflet 10" long, deeply lobed, coarsely toothed; alternate, petiole base fringed. Habitat: rich wet woods. FL: June–July.

SWEET CICELY
Osmorhiza claytonii

CARROT FAMILY   APIACEAE

Perennial, 2' tall, bearing small
flat-topped umbels of small
fragrant white flowers at
top and in axils of leafy
stem with pinnate leaves.
Male and complete
flowers in umbel. Flower
$1/16$" wide, corolla 5-lobed, reflexed.
Calyx none. Stamens 5. Ovary inferior, short
forked style. Fruit $1/2$" linear, black, flat,
pointed, covered with bristles, styles at tip.
Flower stalk with fuzzy white hairs. Leaves 2×
pinnate, leaflets 2" long, petioled or sessile, round-toothed, tip leaflet
deeply lobed. Petiole smooth among leaflets but hairy and reddish at
node. Habitat: moist often rocky woods. FL: May–June.

WATER-PARSNIP
Sium suave

CARROT FAMILY   APIACEAE

Perennial, 5' tall, bearing large
umbels of fragrant white flowers
on top of leafy ribbed smooth
hollow stem. Flower $1/8$" wide,
petals 5. Sepals tiny. Stamens 5.
Ovary inferior, 2 very short
styles. Fruit small round
and ribbed pod.
Umbel of 6 or more
umbellets, all
subtended by short,
reflexed bracts. Lower
leaves 10" long, pinnate, 5–17
narrow leaflets 4" long × $1/4$" wide,
sharply toothed, pointed, sessile. Upper
leaves narrow, 2 basal lobes; alternate. Submerged leaves, if
present, dissected. Habitat: wet meadows, ponds. FL: July–September.

HEMP-DOGBANE
Apocynum cannabinum
<small>DOGBANE FAMILY</small>   <small>APOCYNACEAE</small>

Perennial, 4' tall, bearing
small terminal and axillary
clusters of fragrant nodding
bell-shaped tubular white
flowers on leafy reddish stem with
milky sap. Flower $^{3}/_{16}$" wide, corolla tube
$^{1}/_{8}$" long, 5-lobed, pointed. Calyx cream, 5-
lobed, pointed, shorter than
corolla. Stamens 5, short,
adherent to stigma. Ovary
superior, 2-lobed, with 5 basal
nectaries, 1 style, large stigma. Fruit V-shaped
pair of 2 follicles, seeds with tuft of hair. Leaves 3"
long, oval, pointed, smooth, entire, short-petioled,
partially clasping; opposite. Habitat: sandy beaches; rare. FL: June–
September.

WHORLED MILKWEED
Asclepias verticillata
<small>MILKWEED FAMILY</small>   <small>ASCLEPIADACEAE</small>

Perennial, 20" tall, bearing umbels of
white flowers on stem with narrow
whorled leaves and milky sap. Flower
$^{1}/_{4}$" wide, corolla 5 reflexed white
lobes. Calyx 5, small lobes. Stamens
5, encircle style-stigma and together
form light-green flat gynostegium
at flower center. Five cup-like
hoods equal height of gynostegium.
Five incurved pointed horns are
taller than hoods. Ovary superior, 2.
Fruit 3" long smooth, narrow pods;
seeds flat with silky hairs. Leaves 2" long ×
$^{1}/_{16}$" wide, 3–6 whorled, edges inrolled.
Habitat: moist open woods. FL: June–August.

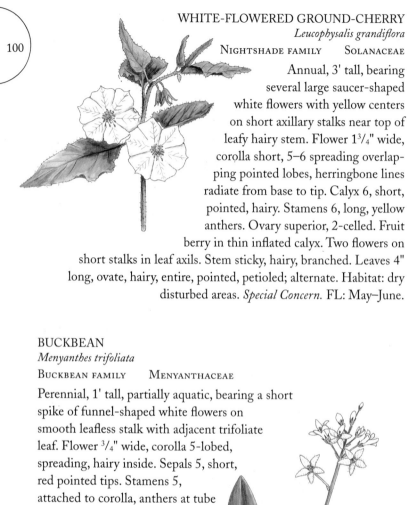

WHITE-FLOWERED GROUND-CHERRY
Leucophysalis grandiflora

NIGHTSHADE FAMILY   SOLANACEAE

Annual, 3' tall, bearing several large saucer-shaped white flowers with yellow centers on short axillary stalks near top of leafy hairy stem. Flower 1³/₄" wide, corolla short, 5–6 spreading overlapping pointed lobes, herringbone lines radiate from base to tip. Calyx 6, short, pointed, hairy. Stamens 6, long, yellow anthers. Ovary superior, 2-celled. Fruit berry in thin inflated calyx. Two flowers on short stalks in leaf axils. Stem sticky, hairy, branched. Leaves 4" long, ovate, hairy, entire, pointed, petioled; alternate. Habitat: dry disturbed areas. *Special Concern.* FL: May–June.

BUCKBEAN
Menyanthes trifoliata

BUCKBEAN FAMILY   MENYANTHACEAE

Perennial, 1' tall, partially aquatic, bearing a short spike of funnel-shaped white flowers on smooth leafless stalk with adjacent trifoliate leaf. Flower ³/₄" wide, corolla 5-lobed, spreading, hairy inside. Sepals 5, short, red pointed tips. Stamens 5, attached to corolla, anthers at tube throat. Ovary superior, round, 1-celled, stigma 2-lobed. Fruit capsule, small shiny seeds. Flower stalk subtended by narrow, pointed bract. Leaves on long petiole from rhizome. Leaflets 3, sessile, smooth, entire, elliptic, fleshy. Habitat: cold bogs, shallow cold water. FL: May–June.

AMERICAN GROMWELL
Lithospermum latifolium
BORAGE FAMILY   BORAGINACEAE

Perennial, 30" tall, bearing single
funnel-shaped white-cream
flower in axils along coiled
branches on leafy downy
stem. Flower $^3/_{16}$" wide,
corolla 5 spreading lobes.
Calyx 5-lobed, finely hairy.
Stamens 5, short. Ovary
superior, short style. Fruit hard
smooth nutlets. Stem branched
above, hairs pointing in one
direction. Leaves 3" long, oval-
lanceolate, sessile, blunt tip, over
$^1/_2$" wide. Habitat: rich woods. FL: May–June.
Gromwell, *L. officinale*, similar but leaves less than $^1/_2$" wide, taper to
point, short-petioled. Habitat: disturbed areas. Introduced.

AMERICAN WATER-HOREHOUND
Lycopus americanus
MINT FAMILY   LAMIACEAE

Perennial, 4' tall, bearing dense clusters of irregular small white flowers
in axils of leafy square stem with coarsely
toothed leaves. Flower $^1/_8$" wide, corolla tube 2-
lipped, upper lip 3-lobed, broader than
lower 2-lobed lip; tube finely hairy.
Calyx tube 4–5 short sharp
teeth. Stamens 4, short pair
in tube, long sterile pair
projects from tube. Ovary
superior, 4-parted, forked
stigma. Fruit 4 gland-dotted nutlets.
Leaves 4" long, lanceolate, taper to short
petiole, lower leaves lobed;
opposite. Upper leaves
toothed, pointed. Habitat: edges
moist woods, ponds. FL: June–September.

CATNIP
Nepeta cataria

MINT FAMILY   LAMIACEAE

Perennial, 3' tall, bearing short spikes of aromatic irregular downy white flowers at top of main stem and branches on leafy hairy square stem. Flower $1/2$" long, tubular corolla 2-lipped, upper lip 2-lobed, lower lip broad and wavy with pink dots, hairs at base of lower lip. Calyx tube irregular, weakly 2-lobed, 5-toothed. Stamens 4 slightly above corolla. Ovary superior, 4-lobed. Fruit nutlets. Leaves $2^{1}/_{2}$" long, ovate, toothed, short-petioled; opposite. Habitat: pastures, fencerows; introduced. FL: June–September.

ENGLISH PLANTAIN
Plantago lanceolata

PLANTAIN FAMILY   PLANTAGINACEAE

Perennial, 1' tall, bearing small tubular white flowers in short cylindrical spike on leafless stalk with basal leaves. Flower tiny, corolla 4-lobed, short. Calyx 3-lobed. Stamens 4, white anthers, project above corolla. Ovary superior, 1 pistil, long style. Fruit 2-seeded capsule (shown), brown oval seeds sticky. Flower subtended by 1–3 short bracts. Leaves 8" long, in basal rosette, narrow, ribbed, lanceolate, pointed. Habitat: disturbed areas; introduced. FL: May–October. Common P., *P. major*, leaves broad, narrow stalk. Red-stemmed P., *P. rugelii*, like Common P., but larger seeds, red petioles. Habitat: disturbed areas.

TURTLEHEAD
Chelone glabra
F̲ɪɢᴡᴏʀᴛ ꜰᴀᴍɪʟʏ   Sᴄʀᴏᴘʜᴜʟᴀʀɪᴀᴄᴇᴀᴇ

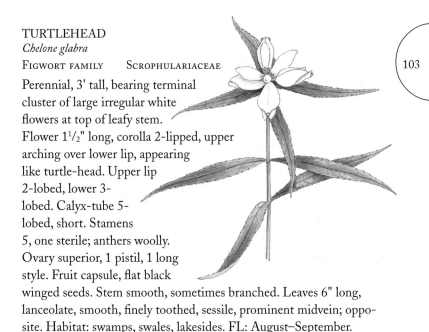

Perennial, 3' tall, bearing terminal
cluster of large irregular white
flowers at top of leafy stem.
Flower 1$^1/_2$" long, corolla 2-lipped, upper
arching over lower lip, appearing
like turtle-head. Upper lip
2-lobed, lower 3-
lobed. Calyx-tube 5-
lobed, short. Stamens
5, one sterile; anthers woolly.
Ovary superior, 1 pistil, 1 long
style. Fruit capsule, flat black
winged seeds. Stem smooth, sometimes branched. Leaves 6" long,
lanceolate, smooth, finely toothed, sessile, prominent midvein; opposite. Habitat: swamps, swales, lakesides. FL: August–September.

COW-WHEAT
Melampyrum lineare
F̲ɪɢᴡᴏʀᴛ ꜰᴀᴍɪʟʏ   Sᴄʀᴏᴘʜᴜʟᴀʀɪᴀᴄᴇᴀᴇ

Annual, 1' tall, bearing small irregular
tubular cream-white flowers in
axils at top of leafy stem.
Flower $^1/_2$" long, corolla
tube 2-lipped, upper lip
cream, arched, 2-lobed
and lower lip yellow,
lobed. Calyx tube 4–5
lobes, short. Stamens 4
fertile, and 1 sterile
stamen tipped with tuft of
hair. Ovary superior, 2-
celled, style below upper lip. Fruit
4-seeded capsule. Flower bracts short,
sometimes toothed. Stem smooth, often
branched. Leaves 4" long, narrow-lanceolate, short-petioled; opposite.
Habitat: dry sandy woods, rocky ledges. FL: June–August.

CATCHWEED
Galium aparine

MADDER FAMILY   RUBIACEAE

Reclining annual, 3' long, bearing tiny white
flowers on short stalks in axils of 8 whorled
leaves on a leafy bristly stem. Flower ⅛" wide,
corolla 4 oval pointed lobes. Sepals none.
Stamens 4. Ovary inferior, 1 pistil, short styles,
globose stigmas. Fruit 2-lobed dry capsule
with hooked hairs that stick to fur, clothing.
Stem 4-angled, recurved hairs on angles, weak
stem. Leaves 2" long, linear, taper to stem, stiff-
haired, sessile; 8-whorled. Habitat: moist
woods, meadows, thickets; introduced. FL:
May–July.

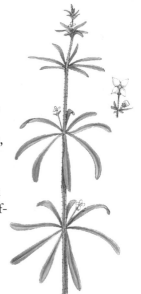

LABRADOR-BEDSTRAW
Galium labradoricum

MADDER FAMILY   RUBIACEAE

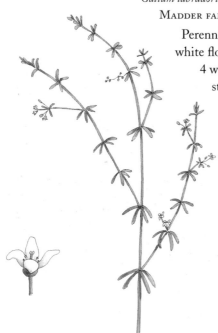

Perennial, 1' tall, bearing several tiny
white flowers on short stalk in axils of
4 whorled leaves on leafy smooth
stem. Flower 1/16" wide, corolla
4 oval lobes. Sepals none.
Stamens 4, brown anthers.
Ovary inferior, 2 short
styles, globose stigma.
Fruit 2-lobed smooth dry
capsule. Flowers 2–4 on
axillary stalk. Stem
mostly erect, smooth but
fine hairs at nodes. Leaves
⅝" long, linear, hang
downward or reflexed, some
marginal hairs, sessile; 4-
whorled. Habitat: fens, bogs.
FL: June–July.

PARTRIDGE-BERRY
Mitchella repens

Creeping perennial, 1' long, bearing pairs of hairy funnel-shaped white flowers on leafy stem. Flowers of 2 types on different plants: (1) short filaments, long styles, or (2) long filaments, short styles. Flower $^3/_8$" long, corolla 4-lobed, spreading, hairy. Corolla tube of both flowers emerge from common cup of 2 fused calyxes. Stamens 4, either short or long filaments. Ovary inferior, 1 pistil with either short or long style, X-shaped stigma. Fruit red berry. Leaves $^5/_8$" long, round-oval, smooth, entire, petioled, evergreen; opposite. Habitat: moist rich woods. FL: May–July.

MAPLE-LEAVED VIBURNUM
Viburnum acerifolium

HONEYSUCKLE FAMILY   CAPRIFOLIACEAE

Perennial, 6' tall, bearing flat-topped cluster of fragrant bell-shaped white flowers at top of leafy stem. Flower $^1/_4$" wide, corolla 5-lobed, spreading. Calyx reduced. Stamens 5, extend above corolla lobes. Ovary inferior, short 3-lobed stigma-style. Fruit purple drupe, hard grooved seed. Marginal flowers same as central flowers. Stem woody, shrubby. Leaves 5" long, maple-like 3-lobed, toothed, petioled, stellate hairs and black dots on underside; opposite. Habitat: medium woods. FL: May–August.

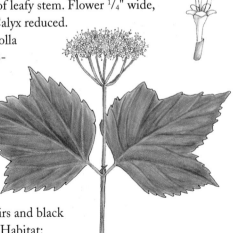

MARSH VALERIAN
Valeriana uliginosa
VALERIAN FAMILY   VALERIANACEAE

Perennial, 4' tall, bearing small clusters of tiny white flowers at top of smooth stem with pinnate leaves. Flower $3/16$" long, corolla with 5 spatulate spreading lobes. Calyx white with green midvein. Stamens 3, white anthers. Ovary inferior, 1 pistil, 1 style. Fruit elongated achene. Cluster subtended by bract. Basal leaves 2' long, deeply pinnate, 5–8 pairs coarse-toothed leaflets, lower region of petiole U-shaped in cross-section, leaflets attached at edge of U. Upper leaflets narrow-elliptic, entire, tip lobe toothed. Habitat: swamps. *Threatened.* FL: June–July.

YARROW
Achillea millefolium
DAISY FAMILY   ASTERACEAE

Perennial, 3' tall, bearing flat-topped cluster of white flower-heads at top of aromatic hairy stem with fern-like leaves. Head $1/4$" wide, 4–6 ray, and tiny disk flowers. Ray flower corolla broad, spreading. Disk flower corolla tubular, 5-toothed, cream-colored. Calyx pappus absent. Stamens 5, encircle style. Ovary inferior, 1 pistil, style lobes short, flat. Fruit flattened smooth achene; pappus absent. Involucral bracts scaly. Leaves 6" long, overall linear shape and much dissected, sessile; alternate. Habitat: roadsides, disturbed areas; introduced. FL: June–September.

PEARLY EVERLASTING
Anaphalis margaritacea

DAISY FAMILY   ASTERACEAE

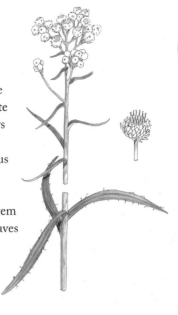

Perennial, 3' tall, bearing a cluster of
many globose flower-heads with white
petal-like bracts at top of cottony white
leafy stem. Head ¹/₄" wide, disk flowers
only. Female and male plants. Female
and male corollas tubular. Calyx pappus
of fine bristles. Stamens 5, yellow
anthers. Ovary inferior, 1 pistil, style
forked. Fruit warty achene, pappus. Stem
cottony white, runners form mats. Leaves
5" long, narrow, sessile, pointed, long
hairs on margin, gray-green above,
cottony white below; alternate.
Habitat: dry fields, disturbed areas.
FL: July–September.

FIELD-PUSSYTOES
Antennaria neglecta

DAISY FAMILY   ASTERACEAE

Perennial, 18" tall, bearing small dense cluster of
white flower-heads on woolly white leafy stem.
Head ³/₈" wide, all disk flowers; resemble cat's
paw. Corolla thin, tubular. Calyx pappus of
bristles. Stamens 5. Ovary inferior, 1
pistil, forked style. Fruit brown resin-
dotted achene, pappus. Involucral
bracts chaffy. Separate male and
female plants; females frequent (shown).
Stem leaves ³/₄" long, linear, woolly,
entire sessile; alternate. Basal
leaves 1" long, broad, sessile, 1
major vein. Habitat: roadsides, fields. FL: April–June.
Plantain-p., *A. plantaginifolia*,
smaller, leaves 3-veined. Habitat:
woodland openings, dry areas.

NORTHERN BOG-ASTER
Aster borealis

Perennial, 4' tall, bearing several white or light-blue flower-heads at top of leafy often red stem with narrow leaves. Heads, few, 3/4" wide, 20–50 ray, and yellow disk flowers. Calyx pappus of hairs. Ovary inferior, 1 pistil, forked style. Fruit smooth achene, pappus. Involucral bracts smooth, slender, overlapping. Narrow bracts on stalks. Leaves 5" long × 1/4" wide, grass-like, margin rough, sessile or base may clasp stem, pointed, smooth but midrib may be hairy on underside; alternate. Habitat: cold bogs, wet meadows. FL: August–October.

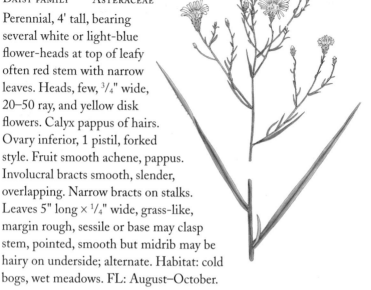

MANY-FLOWERED ASTER
Aster ericoides

DAISY FAMILY   ASTERACEAE

Perennial, 3' tall, bearing many small white flower-heads at top of leafy finely hairy stem with sharp-pointed leaves and bracts. Head 1/4" wide, 8–20 ray, and yellow disk flowers. Calyx pappus of hairs. Ovary inferior, 1 pistil, forked style. Fruit minutely hairy achene, pappus. Involucral bracts large, outer ring green, spiny, marginal hairs, overlapping. Inflorescence bracts many, narrow, sharply pointed. Leaves 1" long × 1/8" wide, linear, most smaller, sessile, entire, finely hairy, prickly to touch; alternate. Lower leaves fall off early. Habitat: dry open meadows. FL: August–October.

EASTERN LINED ASTER
Aster lanceolatus, var. *interior*

DAISY FAMILY   ASTERACEAE

Perennial, 4' tall,
bearing many small
white flower-heads on
branched leafy inflorescence
with lines of hairs on stem.
Head $^3/_8$" wide, 20–40 rays,
and yellow disk flowers. Disk
corolla lobes $^1/_3$ length of
entire corolla. Calyx pappus of hairs.
Ovary inferior, 1 pistil, forked style. Fruit
ribbed finely hairy achene, pappus. Long
leaf below major axillary flower stalks, nearly
as long as stalk. Leaves 5" long × $^1/_4$" wide, linear, entire, broad base,
sessile; alternate. Habitat: moist low areas. FL: August–October.
Variety *lanceolatus*, similar but larger $^5/_8$" wide heads, $^1/_2$" wide leaves.
Habitat: moist low meadows.

GOBLET-ASTER
Aster lateriflorus

DAISY FAMILY   ASTERACEAE

Perennial, 4' tall, bearing short axillary
stalks of several white flower-heads on
seemingly one-sided inflorescence at top
of leafy finely hairy stem. Head $^1/_2$"
wide, 20–30 ray, and yellow disk
flowers. Calyx pappus of hairs. Ovary
inferior, 1 pistil, forked style. Fruit
ribbed hairy achene with pappus.
Involucral bracts, broad, pointed,
overlapping, green midrib. Leaves 5"
long, lanceolate, faintly toothed,
pointed, smooth but midrib hairy
below. Lower leaves petioled,
upper leaves sessile;
alternate. Habitat: open
woods. FL: August–October.

OX-EYE DAISY
Chrysanthemum leucanthemum

DAISY FAMILY   ASTERACEAE

Perennial, 2' tall, bearing one white and yellow-centered flower-head on smooth leafy stalk with basal leaves. Head 2" wide, 10 or more ray, and many yellow disk flowers; disk depressed at center. Calyx pappus none. Stamens 5. Ovary inferior, 1 pistil, style bilobed, flat with short hairs. Fruit black angular ribbed achene, beaked. Involucral bracts narrow, brown edges. Basal leaves 6" long, petioled, spatulate, pinnately lobed with rounded segments, smooth. Upper leaves pinnately lobed, smaller, clasping; alternate. Habitat: disturbed areas; introduced. FL: May–October.

DUNE-THISTLE
Cirsium pitcheri

DAISY FAMILY   ASTERACEAE

Perennial, 3' tall, bearing several large fragrant white flower-heads at top of gray-green woolly stem with pinnately lobed leaves. Head 1¹/₂" wide, disk flowers only. Corolla white and violet. Calyx pappus of feathery bristles. Stamens 5. Ovary inferior, 1 pistil, lobed style hairy. Fruit flat achene, pappus. Involucre 1" high, bracts weakly spiny, woolly. Leaves deeply cleft into linear lobes, white midvein, weakly spine-tipped, white woolly, winged petiolar base extends down stem; alternate. Habitat: sand dunes Lake Michigan shore. *Threatened.* FL: June–September.

HORSEWEED
Conyza canadensis

<small>DAISY FAMILY</small>   <small>ASTERACEAE</small>

Annual, 4' tall, bearing many small flower-heads with white ray flowers and yellow center on many branches in upper axils on leafy minutely hairy stem. Head $\frac{1}{4}$" wide, 10–25 ray flowers, erect. Disk flowers tubular. Calyx pappus of long bristles, evident around disk flowers. Ovary inferior, 1 pistil, style lobes appendaged. Fruit 1-ribbed achene, pappus. Involucral bracts green in 1 ring, short, narrow, midvein evident, overlapping; reflexed on dry head. Leaves 1" long, narrow, entire, sessile; alternate. Habitat: disturbed areas. FL: May–July.

ROUGH FLEABANE
Erigeron strigosus

<small>DAISY FAMILY</small>   <small>ASTERACEAE</small>

Annual, 30" tall, bearing flower-heads with white rays and yellow center on leafy hairy stem. Head $\frac{1}{2}$" wide, 50–100 ray, and many disk flowers. Disk calyx pappus of bristles. Ovary inferior, style flat, appendaged. Fruit 2-nerved achene, pappus. Involucral bracts shorter than width of disk, overlapping. Basal leaves 4" long, entire. Stem leaves 2" long, entire, sessile: alternate. Habitat: disturbed sunny areas; introduced. FL: June–October.
Annual F., *E. annuus,* similar but spreading hairs on stem, leaves petioled, slightly clasping, toothed. Habitat: disturbed sunny meadows.

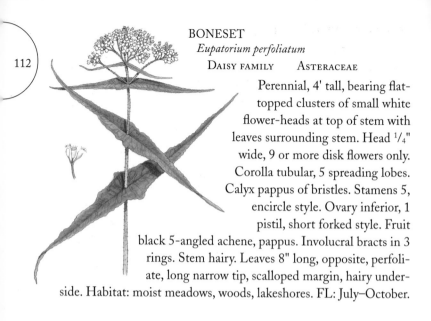

BONESET
Eupatorium perfoliatum
DAISY FAMILY   ASTERACEAE

Perennial, 4' tall, bearing flat-topped clusters of small white flower-heads at top of stem with leaves surrounding stem. Head $1/4$" wide, 9 or more disk flowers only. Corolla tubular, 5 spreading lobes. Calyx pappus of bristles. Stamens 5, encircle style. Ovary inferior, 1 pistil, short forked style. Fruit black 5-angled achene, pappus. Involucral bracts in 3 rings. Stem hairy. Leaves 8" long, opposite, perfoliate, long narrow tip, scalloped margin, hairy underside. Habitat: moist meadows, woods, lakeshores. FL: July–October.

WHITE SNAKEROOT
Eupatorium rugosum
DAISY FAMILY   ASTERACEAE

Perennial, 3' tall, bearing clusters of fuzzy white flower-heads on hairy stem. Head $3/16$" wide, 8–25 disk flowers.
Corolla 5 spreading lobes. Calyx pappus of bristles. Stamens 5, encircle style. Ovary inferior, very long forked style. Fruit smooth 5-angled achene, pappus. Involucral bracts similar length, not overlapping. Leaves 6" long, broadly ovate, often heart-shaped base, pointed, petioled, toothed; opposite. Habitat: rich moist woods. FL: July–October. Late E., *E. serotinum*, similar but red downy stem, narrow ovate leaves, long petiole. Habitat: bottomlands along creeks.

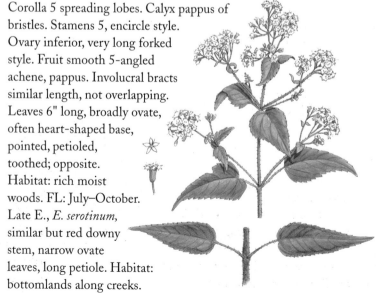

NORTHERN SWEET COLTSFOOT
Petasites frigidus
<small>DAISY FAMILY</small>   <small>ASTERACEAE</small>

Perennial, 18" tall, bearing terminal cluster of
fragrant white flower-heads on thick stem
with many bracts. Heads $^3/_8$" wide,
many ray, and 7–8 disk flowers.
Disk flower corolla star-shaped.
Ray flower broad, tooth-like, tip
lobed, long style. Calyx pappus of
white bristles. Stamens 5, yellow.
Ovary inferior, 1 pistil, forked
style. Fruit linear ribbed achene,
pappus. Involucral bracts red-
tipped in one row. Each head on
short scaly stalk subtended by bract.
Leaves 10" wide, appear after flowering,
long-petioled, triangular to round, deeply palmately lobed, smooth or
finely hairy. Habitat: moist swampy woods. FL: April–July.

RATTLESNAKE-ROOT
Prenanthes alba
<small>DAISY FAMILY</small>   <small>ASTERACEAE</small>

Perennial, 5' tall, bearing spike of
several fragrant nodding white flower-
heads at top of leafy smooth stem with
milky sap. Heads $^5/_8$" wide, 8–12 ray
flowers only. Corolla square-
tipped, 5-notched. Calyx
pappus of brown bristles.
Stamens 5, pink, prominent,
encircle long style. Ovary
inferior, 1 pistil. Fruit curved
ribbed reddish achene, pappus.
Involucral bracts 8 long and several short,
smooth. Lower leaves 8" long, triangular,
petioled, lobed, alternate; upper smaller,
triangular-lanceolate. Habitat: rich woods,
thickets. FL: August–September.

WATER-PLANTAIN
Alisma triviale

WATER-PLANTAIN FAMILY ALISMATACEAE

Perennial, 3' tall, bearing small white
flowers at tips of flower stalks
arranged in whorls on much
branched leafless stem adjacent
to long-petioled leaves.
Flower 3/16" wide, petals 3,
cup-like. Sepals 3, short,
green-striped with step-like
margin, cup-shaped. Sta-
mens 6, short, cream pollen.
Ovary superior, usually 10, in
one whorl (shown). Fruit achene. Small
bracts below each whorl of flower stalks on
floral axis. Leaves 12" long, basal, long-petioled, oval, pointed, parallel-
veined. Habitat: marshes, lake edges. FL: June–September.

COMMON ARROWHEAD
Sagittaria latifolia

WATER-PLANTAIN FAMILY ALISMATACEAE

Perennial, 2' tall, bearing fragrant white
flowers in whorls of 3 on leafless
stalk adjacent to arrow-shaped
leaves. Flower 1" wide, 3 broad
petals, ruffled margins. Sepals
3, green, reflexed. Stamens
many, yellow. Ovary superior,
many pistils each with style.
Fruit burred ball of winged
achenes with oblique beak.
Lower flowers female on long
stalks; upper, male on short stalks.
Bracts below whorls of flowers. Leaves
basal, long petiole, blades of different
widths. Submerged leaves, if present,
ribbon-like. Habitat: shallow water, wet
banks. FL: June–August.

WILD-CELERY
Vallisneria americana
FROG'S-BIT FAMILY   HYDROCHARITACEAE

Perennial aquatic, 4' tall,
bearing solitary whitish
female flower on long slender
leafless stalk at water surface.
Plants male or female
(shown). Female: flower $1/4$"
wide; petals 3, spreading, at
top of tubular spathe; sepals
3, erect; stigmas 3, broad;
ovary inferior; fruit few-seeded capsule.
Male: flower $1/16$" wide; released from plant
and floats to water surface where it opens to
expose pollen. Fertilized female retracts ovary
under water for seed maturation. Leaves $1/2$"
wide, linear to 4' long, tips often floating. Habitat:
quiet flowing water. FL: July–October.

WATER-ARUM
Calla palustris
ARUM FAMILY   ARACEAE

Perennial, 1' above water, bearing broad white pointed spathe around
yellow-green flower-covered spadix at top of leafless stalk. Flowers
many, tiny green and usually perfect, or with female at base and male
near top of spadix. Petals, sepals none. Stamens 6. Ovary
superior, pistil of a flower
imbedded in spadix,
stubby
stigma-style.
Fruit red berry,
many on spadix.
Spathe $1^1/2$" wide, blade-
like. Stem rhizome produces
erect, long-petioled heart-shaped,
glossy 6" leaves, smooth, entire.
Habitat: cold marshes, shallow
water. FL: May–August.

WILD LEEK
Allium tricoccum
LILY FAMILY   LILIACEAE

Perennial, 18" tall, bearing round umbel of white flowers each with green center on leafless stem; leaves absent. Flower $1/4$" wide, perianth cup-like, 6 tepals, 3 spread outwardly and 3 turn inwardly. Stamens 6, short. Ovary superior, green, short style. Fruit 3-lobed capsule; seeds black. Two narrow 1" long bracts below umbel of 20 or more flowers. Leaves 1' long, 1–3" wide, flat, lanceolate, parallel veins, smooth, entire, narrow to reddish petiolar base, onion odor. Leaves appear early in spring, often wither before flower stalk appears. Habitat: rich woods. FL: June–July.

WILD ASPARAGUS
Asparagus officinale
LILY FAMILY   LILIACEAE

Perennial, 6' tall, bearing individual nodding bell-shaped cream flowers on slender stalks in leaf axils on stem with feather-like leaves. Flower $3/8$" long, perianth 6 spreading partly fused tepals. Stamens 6, short, yellow anthers. Ovary superior, 3 short stigma-styles. Fruit $3/8$" red berry. Green stem, much branched. Leaves modified into short scales below green thread-like branches. Habitat: fields, road-sides; introduced. FL: June–July.

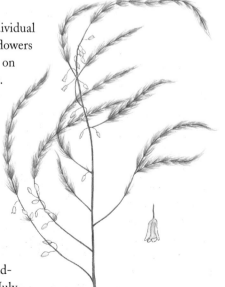

CANADA MAYFLOWER
Maianthemum canadense

LILY FAMILY   LILIACEAE

Perennial, 8" tall, bearing dense terminal cluster of star-like white flowers at top of zigzag stem with 1–3 leaves. Flower $1/8$" long, perianth 4 spreading tepals. Stamens 4, cream. Ovary superior, 2-celled, short slightly lobed style, globose stigma. Fruit globose white berry speckled red, matures to red. Leaves 2" long, usually 2–3, heart-shaped base, shiny, upper side smooth, entire, parallel venation, sessile or short-petioled, underside either woolly (one variety) or smooth (another variety); alternate. Habitat: medium, moist woods. FL: May–June.

FALSE SOLOMON'S SEAL
Smilacina racemosa

LILY FAMILY   LILIACEAE

Perennial, 3' tall, bearing terminal pyramidal cluster of creamy white flowers at tip of arching leafy zigzag stem. Flower $1/4$" wide, perianth 6 tepals, somewhat shorter than the 6 stamens; tepals quickly fall off during flowering. Ovary superior, 3-celled, short stigma-style. Fruit red berry dotted purple. Leaves 6" long, flat, elliptic, pointed, red nodes, hairy along veins on underside, sessile, parallel veins; alternate. Habitat: medium woods, clearings. FL: May–July.

STARRY FALSE SOLOMON'S SEAL
Smilacina stellata

LILY FAMILY   LILIACEAE

Perennial, 2' tall, bearing a short
terminal row of star-shaped
white flowers on arching
leafy zigzag stem. Flower
$1/4$" wide, perianth 6
tepals, narrow, pointed.
Stamens 6, yellow
anthers. Ovary superior,
short style, lobed stigma.
Fruit round black-striped
greenish berry, turns maroon. Leaves
4" long, lanceolate, somewhat folded
along center, sessile to clasping, entire,
minutely hairy underside; alternate. Habitat:
dry wood edges, sandy areas, lake shores. FL: May–August.
Three-leaved False S. S., *S. trifolia,* similar but only a few
flowers with 3 leaves. Habitat: bogs.

FALSE ASPHODEL
Tofieldia glutinosa

LILY FAMILY   LILIACEAE

Perennial, 2' tall, bearing terminal spike
of many small white flowers on tall
sticky leafless stalk with basal
grass-like leaves. Flower $3/16$" long,
perianth 6 smooth tepals. Anthers
6, short, equal tepal length. Ovary
superior, 3-celled, 3 styles. Fruit
oblong capsule, persistent perianth
at base; seed with thread-like
appendage longer than seed body.
Bract below flower. Stalk sticky, black
glands, bract above midpoint. Leaves
18" long × $1/4$" wide, several, linear, 2-
ranked. Habitat: cool beaches. *Threatened.*
FL: July.

NODDING TRILLIUM
Trillium cernuum

LILY FAMILY   LILIACEAE

Perennial, 18" tall, bearing
solitary large white flower
on long nodding stalk with
flower located under whorl
of 3 leaves. Flower 2" wide,
6 tepals: inner 3 white,
widely spread, oval, pointed;
outer 3 light green, shorter than white
tepals. Stamens 6, pink anthers, purple
pollen. Ovary superior, 3-lobed, 3 short
stigmas. Fruit 3-lobed many-seeded
berry. Leaves 4" long, wide, ovate, wavy
margin, pointed, smooth, short-
petioled; whorled. Habitat: wet
woods, swamps. FL: April–July.

BIG WHITE TRILLIUM
Trillium grandiflorum

LILY FAMILY   LILIACEAE

Perennial, 18" tall, bearing solitary large waxy white
flower on long erect stalk above single
whorl of 3 leaves. Flower 4" wide,
perianth of 6 tepals, 3 white large,
round pointed inner tepals with
wavy margin; 3 green outer,
narrow, pointed tepals. Stamens
6, yellow anthers. Ovary
superior, 6 wing-like angles,
3 short stigmas. Fruit 3-
celled berry. Flowers
turn pink with age.
Leaves 6" long,
oval, wavy
margin, pointed,
sessile; whorled. Habitat: rich moist
woods. FL: April–June.

DEATH-CAMAS
Zigadenus elegans
LILY FAMILY   LILIACEAE

Perennial, 3' tall, bearing loose spike of many star-like greenish-white flowers on leafless stalk with adjacent grass-like leaves. Flower $1/2$" wide, perianth of 6 tepals, 3 tepals with purple tips, all tepals with yellow-green horizontal stripe near base forming a central ring. Each tepal with 2 basal glands. Stamens 6, dark purple anthers. Ovary superior, 3-lobed, 3 styles. Fruit 3-lobed capsule, styles persistent. Each flower subtended by purple bract. Leaves basal, $2^1/_2$' long, narrow, pointed. Poisonous if eaten. Habitat: calcareous swamps, sandy beaches. FL: June–August.

SHOWY LADY-SLIPPER
Cypripedium reginae
ORCHID FAMILY   ORCHIDACEAE

Perennial, 3' tall, bearing large fragrant irregular white flowers with slipper-like pink pouch on leafy stem. Flower 4" wide, petals 3, lower one inflated into pink-streaked slipper; lateral 2 petals spreading. Sepals 3, upper 1 forms hood over lip, lateral 2 fused into lower sepal. Stamen anthers 2, fused with style to form column that projects into pouch; staminode covers stigma. Ovary inferior. Fruit capsule. Bract below flower. Leaves 10" long, finely hairy, elliptic-elongate, ribbed, base clasps stem; alternate. Habitat: mossy bogs, swamps. *Special Concern.* FL: May–August.

LESSER RATTLESNAKE-PLANTAIN
Goodyera repens
ORCHID FAMILY   ORCHIDACEAE

Perennial, 1' tall, bearing one-sided spike of
irregular white flowers with sac-like lip on
leafless stalk with checkered basal leaves.
Flower $^3/_8$" long, petals 3, upper 2 fused
to upper sepal to form hood, lower
petal forms greenish basal sac with
downward projecting lip. Two remaining white sepals point downward.
Stamen anthers fused with style to form
column tipped with stigma. Ovary inferior, 1-
celled. Fruit capsule. Bract below flower;
scales on stalk. Leaves 2" long, broad-
ovate. Habitat: under hemlocks,
spruces. FL: July–August.

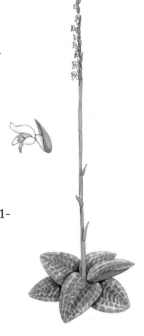

NODDING LADIES'-TRESSES
Spiranthes cernua
ORCHID FAMILY   ORCHIDACEAE

Perennial, 2' tall, bearing spike of fragrant
irregular nodding white flowers in 3 spirals
on leafless stalk. Flower $^1/_4$" wide, petals 3,
lower petal forms tongue-like lip with
wavy margin and 2 basal lobes; 2 lateral
petals curve upward and appress sepal
between them. Sepals 3, upper
between petals and 2 laterals project
forward, hairy. Stamen anthers 2, fused
with style and pollen gland to form
column over throat. Ovary inferior.
Fruit capsule. Hairy bract below
flower; several bracts along stalk.
Basal leaves keeled, 4" long, usually
2, grass-like. Habitat: wet sandy
meadows. FL: August–September.

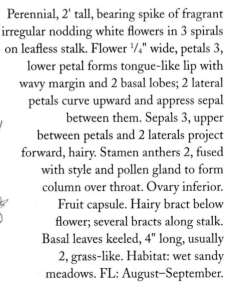

HOODED LADIES'-TRESSES
Spiranthes romanzoffiana
ORCHID FAMILY   ORCHIDACEAE

Perennial, 2' tall, bearing spike of fragrant irregular white flowers spirally arranged on leafless stalk. Flower $1/4$" wide. Petals 3, long lower petal forms lip with wavy margin, reflexed; 2 lateral petals project forward. Sepals 3, upper sepal forms broad hood appressed to and covering 2 lateral petals; 2 lateral sepals project forward. Stamen anthers 2, fused with style and pollen gland into column in throat. Ovary inferior. Fruit capsule. Hairy bract below flower; several bracts along stalk. Leaves basal, 4" long, 1–2, grass-like. Habitat: bogs, wet meadows. FL: July–August.

Pink

FLOWERS

ALLEGHENY-VINE
Adlumia fungosa
FUMITORY FAMILY   FUMARIACEAE

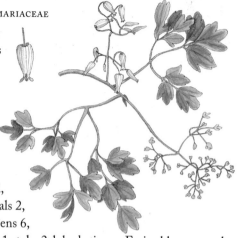

Biennial vine, 10' long, bearing drooping clusters of small elongated irregular pink flowers on axillary stalks on leafy stem. Flower $1/2$" long, petals 4, papery, outer 2 inflated into elongated balloon with flared lips, and small inner 2 hooded, arched over stamens. Sepals 2, small, quickly shed. Stamens 6, greenish. Ovary superior, 1 style, 2-lobed stigma. Fruit oblong capsule. Stem, leaves smooth, exude milky sap. Leaves 10" long, 2–3× pinnate, leaflets lobed, long petiole wraps around host to support vine; alternate. Habitat: wooded slopes. *Special Concern.* FL: June–August.

WATER-HEMP
Amaranthus tuberculatus
AMARANTH FAMILY   AMARANTHACEAE

Annual, 3' tall, bearing clusters of small fuzzy pink flowers at top and in axils of branched leafy stem. Flower $1/8$" wide, sparkling, crystals on its surface. Separate female and male plants; no petals. Male: sepals 5, unequal; stamens 5; short greenish-pink bracts below flower. Female: no sepals, petals or stamens; superior ovary, 3 glistening pink, long, hairy, pointed stigmas (shown). Pink bracts below flower. Fruit inflated utricle, brown seeds. Leaves 3" long, lanceolate or oval, smooth, entire, petioled; alternate. Leaves among flower clusters. Habitat: stream banks. FL: July–October.

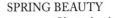

SPRING BEAUTY
Claytonia virginica

PURSLANE FAMILY   PORTULACACEAE

Perennial, 8" tall, bearing several pink flowers in terminal cluster between pair of opposite leaves. Flower ½" wide, petals 5, with dark pink stripes. Sepals 2, small. Stamens 5, white. Ovary superior, 1 pistil, 3-lobed style. Fruit capsule, 3–6 seeds. Stem short, reddish, weak, from tuber. Leaves 2" long, opposite, linear, more than 8× longer than wide, sessile or short petiole; opposite. Habitat: rich moist woods. FL: April–May. *C. caroliniana*, leaves oblong, less than 8× as long as wide, distinct petiole. Habitat: rich moist rocky woods.

BOUNCING BET
Saponaria officinalis
PINK FAMILY   CARYOPHYLLACEAE

Perennial, 30" tall, bearing cluster of fragrant large flask-shaped pale pink flowers at top of leafy stem. Flower 1" wide, petals 5, long, spreading, claw, inturned and notched tips. Calyx tube ¾" long, 5-toothed, cylindrical, smooth. Stamens 5, protrude from throat. Ovary superior, 2 styles often protruding. Fruit capsule. Stem smooth, often wine-colored. Leaves 2" long, oblong-lanceolate, wavy margin, smooth, sessile, conspicuous veins; opposite. Habitat: disturbed areas along lakes; introduced. FL: July–September.

WATER SMARTWEED
Polygonum amphibium
SMARTWEED FAMILY   POLYGONACEAE

Perennial, 3' long, bearing dense
spikes of small pink flowers on
stalk on leafy stem.
Flower ⅛" wide, 5
colored tepals. Stamens
3–8. Ovary superior, 2 styles. Two
flower types: (1) short stamens,
ovary with long styles; (2) long
stamens, ovary with short styles
(shown). Fruit achene. Aquatic
and land varieties. Aquatic var. *stipula-
ceum*, emergent spike; leaves floating, 6" narrow-
oval, round tip, smooth, entire, petioled, fringed
stipule. Habitat: shallow water. Land var. *emersum*, erect
stem with spike; leaves slender, pointed, hairy, glandular petiole.
Habitat: wet shores. FL: July–September.

NODDING SMARTWEED
Polygonum lapathifolium
SMARTWEED FAMILY   POLYGONACEAE

Annual, 3' tall, bearing nodding cylindrical spikes
of pink flowers at top of branched leafy
stem. Flower ⅛" wide, 4–5 tepals,
smooth, fused at base, outer tepals
strongly veined ending in anchor fork.
Stamens few, short. Ovary superior, 1 pistil,
2 styles. Fruit dark achenes, both sides flat.
Smooth dark-pink sheath below flower.
Leaves 5" long, narrow-lanceolate, entire, petioled
base clasps stem, red
midrib widest at leaf base,
stipule toothed but not fringed,
finely hairy underside; alternate.
Habitat: moist sandy beaches. FL:
June–September.

PENNSYLVANIA SMARTWEED
Polygonum pensylvanicum
SMARTWEED FAMILY   POLYGONACEAE

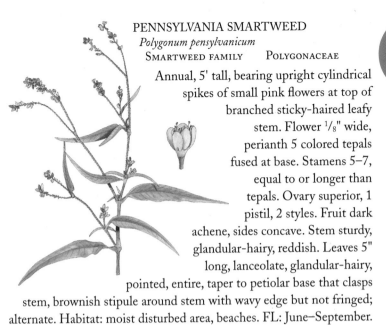

Annual, 5' tall, bearing upright cylindrical spikes of small pink flowers at top of branched sticky-haired leafy stem. Flower $1/8$" wide, perianth 5 colored tepals fused at base. Stamens 5–7, equal to or longer than tepals. Ovary superior, 1 pistil, 2 styles. Fruit dark achene, sides concave. Stem sturdy, glandular-hairy, reddish. Leaves 5" long, lanceolate, glandular-hairy, pointed, entire, taper to petiolar base that clasps stem, brownish stipule around stem with wavy edge but not fringed; alternate. Habitat: moist disturbed area, beaches. FL: June–September.

LADY'S THUMB
Polygonum persicaria
SMARTWEED FAMILY   POLYGONACEAE

Annual, 2' tall, bearing slightly nodding cylindrical spikes of small pink flowers at top of branched leafy stem with spotted leaves. Flower $1/8$" wide, perianth 5 colored tepals, smooth, fused at base. Stamens 8, slender short filaments. Ovary superior, 1 pistil, 2 styles. Fruit black achene. Fringed sheath below flower. Stem smooth, reddish, sprawling. Leaves 5" long, with large red-green central blotch, narrow-elongated, pointed, entire, base clasps stem, stipule fringed with bristles; alternate. Habitat: moist disturbed areas, roadsides; introduced. FL: June–October.

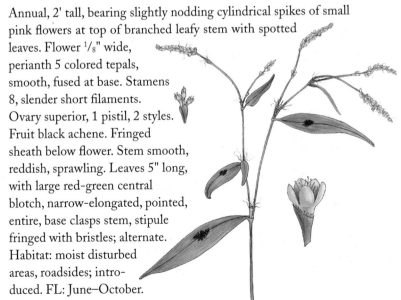

COMMON MALLOW
Malva neglecta
128 MALLOW FAMILY   MALVACEAE

Low growing biennial, 3'
long, bearing small
clusters of pale pink
striped flowers with
notched petals in axils
of leafy stem. Flower $^3/_4$" wide,
petals 5, notched, spreading. Sepals
5, partly united, subtended by 3 narrow
bracts. Stamens many, united into tube
around styles. Ovary superior, 5 to many-
celled. Fruit flat pie-shaped ring of many
seeds with irregular surface, minutely hairy.
Leaves 1$^1/_2$" wide, round, scalloped margins, coarse, toothed,
prominent veins, petiole longer than blade; alternate. Habitat: disturbed areas, drought-resistant plant; introduced. FL: April–October.

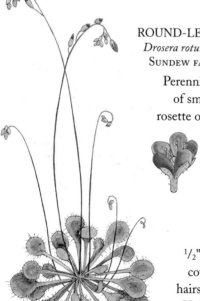

ROUND-LEAVED SUNDEW
Drosera rotundifolia
SUNDEW FAMILY   DROSERACEAE

Perennial, 8" tall, bearing nodding cluster
of small pink flowers on red stalk above
rosette of leaves. Flower $^1/_4$" wide, petals 5,
oval. Sepals 5, covered with sticky
white hairs, shorter than petals.
Stamens 5, shorter than petals.
Ovary superior, 1 pistil, 3 styles.
Fruit capsule, small brown seeds.
Flower cluster one-sided; tips of
unopened buds pink. Leaf
$^1/_2$" wide, round, long-petioled. Blade
covered with red sticky gland-tipped
hairs that trap and digest small insects.
Habitat: peat hummocks in bogs. FL:
June–August.

CUT-LEAVED TOOTHWORT
Cardamine concatenata
MUSTARD FAMILY   BRASSICACEAE

Perennial, 1' tall, bearing terminal cluster of large pink-white or lavender flowers on stalk above whorl of deeply cleft leaves. Flower $3/4$" wide, petals 4, narrow to claw, spatulate, appear opposite, aging to pink. Sepals 4, small. Stamens 6, two short with U-shaped basal gland. Ovary superior, 1 pistil, slender style. Fruit slender round 1" long beaked pod, wrinkled seeds. Upright stem, smooth or slightly woolly, bears whorl of 3 deeply cleft petioled leaves near its midpoint, leaf 4- or 5-parted, lanceolate segments toothed. Habitat: rich moist woods. FL: April–May.

BOG-ROSEMARY
Andromeda glaucophylla
HEATH FAMILY   ERICACEAE

Evergreen perennial, 2' tall, bearing small clusters of drooping bell-shaped pink flowers near top of leafy stems. Flower $1/4$" long, corolla with small 5-lobed mouth. Calyx 5-toothed, short. Stamens 10, short, hairy, anther tipped with spur. Ovary superior, 5-celled, style extends to corolla lip, globose pistil. Fruit $3/8$" globular capsule. Stem shrubby. Leaves 2" long, narrow, margin strongly inrolled, leathery, entire, woolly white below, sessile; alternate. Habitat: bogs. FL: May–June.

COTONEASTER
Cotoneaster divaricatus
HEATH FAMILY   ERICACEAE

Perennial, 4' tall, bearing 1–3 pendant pink bell-shaped flowers on short stalks in axils on arching leafy shrubby stem. Flower ¼" wide, corolla 5-lobed. Calyx 5-lobed. Stamens 10, shorter than corolla. Ovary superior, short stigma-style. Fruit red pome, 2–5 seeds. Leaves ¾" long, elliptical, entire, pointed at both ends, short-petioled, shiny dark green above, light green with short hairs underside; alternate. Habitat: partially shaded wood edges. FL: May–June.

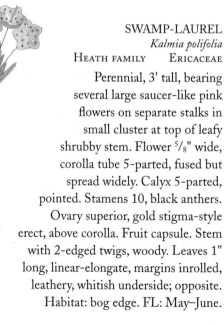

SWAMP-LAUREL
Kalmia polifolia
HEATH FAMILY   ERICACEAE

Perennial, 3' tall, bearing several large saucer-like pink flowers on separate stalks in small cluster at top of leafy shrubby stem. Flower ⅝" wide, corolla tube 5-parted, fused but spread widely. Calyx 5-parted, pointed. Stamens 10, black anthers. Ovary superior, gold stigma-style erect, above corolla. Fruit capsule. Stem with 2-edged twigs, woody. Leaves 1" long, linear-elongate, margins inrolled, leathery, whitish underside; opposite. Habitat: bog edge. FL: May–June.

SMALL CRANBERRY
Vaccinium oxycoccos
<small>HEATH FAMILY</small>   <small>ERICACEAE</small>

Creeping peren-
nial, 6" tall, bearing
typically a single or pair
of nodding pink flowers
on thin axillary stalk on
leafy stem. Flower $^3/_8$" wide,
corolla tube deeply cleft 4-
lobed, linear,
reflexed. Calyx 4-
lobed, small.
Stamens 10, long,
form conical erect cluster. Ovary
superior, dark pink with white
stripes, style longer than stamens.
Fruit round pink-red berry. Flower stalk with 2 bracts near midpoint.
Leaves $^3/_8$" long, flat or with margins curled under, oblong, glossy top,
whitish underside; alternate. Habitat: bogs. FL: June–July.

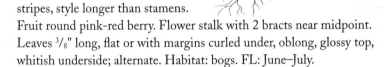

PIPSISSEWA
Chimaphila umbellata
<small>SHINLEAF FAMILY</small>   <small>PYROLACEAE</small>

Perennial, 10" tall, bearing several nodding
waxy pale pink flowers at top of stalk on
stem with whorled leaves. Flower $^5/_8$" wide,
petals 5, round, spreading or reflexed.
Sepals 5, small, reflexed. Stamens 10, short, yellow anthers.
Ovary superior, style short,
globose stigma. Fruit brown
capsule. Stem creeping, with
erect brownish flower stems.
Leaves 2" long, broadest above middle,
shiny bluish-green, lanceolate, petioled,
toothed; whorled. Habitat: rich dry sandy
woods. FL: June–August.

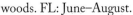

PINK SHINLEAF
Pyrola asarifolia
SHINLEAF FAMILY   PYROLACEAE

Perennial, 1' tall, bearing elongated cluster of waxy nodding fragrant pink flowers on smooth leafless stalk. Flower $^3/_8$" wide, petals 5, oblong, erect, tips darker pink than base. Sepals 5, pointed, long, pale green. Stamens 10, upcurved, shorter than petals; anthers red. Ovary superior, 5-celled; style curved, longer than petals. Fruit capsule. Basal leaves 2" wide, round, leathery, small-toothed, distinct venation, petiole as long or longer than blade. Small scale leaf midway along stalk. Habitat: moist woods, bogs. FL: July–August.

BIRD'S-EYE PRIMROSE
Primula mistassinica
PRIMROSE FAMILY   PRIMULACEAE

Perennial, 8" tall, bearing umbel of several erect pink flowers with yellow centers at tip of leafless stalk with basal rosette of leaves. Flower $^1/_2$" wide, corolla tubular, 5 spreading notched lobes, dark throat surrounded with inner yellow and outer white bands. Calyx 5-lobed, persistent. Stamens 5, yellow. Ovary superior, 1 pistil, 1 long style, globose stigma. Fruit beaked elongated capsule. Several bracts $^3/_{16}$" long below umbel. Leaves 2$^1/_2$" long, spatulate, narrow to base, toothed, hairy underside. Habitat: moist sandy or rocky soil. *Special Concern.* FL: May–July.

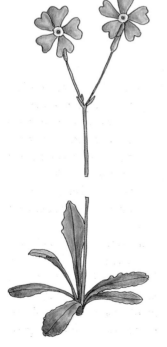

DITCH-STONECROP
Penthorum sedoides
<small>SAXIFRAGE FAMILY　　SAXIFRAGACEAE</small>

Perennial, 2' tall, bearing dusty pink flowers along cluster of short branches at top of leafy stem. Flower ¼" wide, petals none. Sepals 5–6, erect, oblong, pointed. Stamens 10. Ovary superior, 5–6 pink pistils united at their base, appear pointed with short styles (left). Fruit angled follicle, brown seeds. Flowers tend to cluster along one side of branch. Stem simple or branched, smooth below but hairy-glandular along inflorescence. Leaves 4" long, lanceolate, smooth, fine-toothed, short-petioled; alternate. Habitat: creeksides. FL: July–August.

SMOOTH ROSE
Rosa blanda
<small>ROSE FAMILY　　ROSACEAE</small>

Perennial, 3' tall, bearing 1 or few large fragrant pink flowers on separate axillary stalk near top of smooth branched stems with pinnate leaves. Flower 3" wide, petals 5, wider than long. Sepals 5, erect, persistent on fruit. Stamens many, yellow. Ovary inferior, cup-like, ovaries in smooth cup. Fruit achenes in red cup (hip). Leaves alternate, 5–7 pinnate, leaflets 1" long, coarsely toothed especially above middle. Smooth wide stipule. Habitat: sandy area, shores. FL: June–August.

RUGOSE ROSE
Rosa rugosa
ROSE FAMILY   ROSACEAE

Perennial, 6' tall, bearing large fragrant single or double pink flower at shoot tip on very bristly stem with wrinkly pinnate leaves. Flower $3^{1}/_{2}$" wide, petals 5 or many. Sepals 5, erect, hairy. Stamens many, yellow. Cup-like hypanthium smooth. Fruit achenes inside dark red cup (hip). Flower stalk with basal bract, glandular, bristly. Stem bristles longest near nodes. Leaves 7-9 pinnate, petiole bristly, stipules leaf-like, pointed; alternate. Leaflets dark green, elliptical, 1" long, minutely hairy underside. Habitat: sandy areas, roadsides; introduced. FL: June–August.

SWAMP-ROSE
Rosa palustris
ROSE FAMILY   ROSACEAE

Perennial, 6' tall, bearing 1 or few large fragrant pink flowers on separate short stalks on spiny stem with pinnate leaves. Flower 2" wide, petals 5, often wavy. Sepals 5, entire, glandular, soon dropping from young fruit. Stamens many, in central ring. Ovary inferior, cup-like, pistils in glandular cup. Fruit achenes in red cup (hip). Stalk subtended by bract. Stem branched, recurved hooked spines at node; internodes spineless. Leaflets 5–7, each $^{1}/_{4}$" long, finely toothed, downy midrib on underside, stipule very narrow; alternate. Habitat: marshes, swamps. FL: June–August.

EVERLASTING PEA
Lathyrus latifolius
PEA FAMILY   FABACEAE

Trailing perennial, 6' long,
bearing small cluster of large irregular
rose flowers on long stalk in axils on
winged leafy smooth stem
with tendrils. Flower 1" wide,
5 clawed petals, upper petal
broad and notched, 2 dark-pink
lateral petals are small and
incurved, 2 lower green petals are fused to form
keel. Calyx 5-toothed, somewhat unequal.
Stamens 10, in 2 groups. Ovary superior, style
flat, hairy. Fruit smooth green pod (shown).
Leaves 3-parted, forming 2 narrow leaflets and tendril;
stipule narrow with basal lobes. Habitat: disturbed areas;
introduced. FL: June–September.

MERMAID-WEED
Proserpinaca palustris
WATER-MILFOIL FAMILY   HALORAGACEAE

Perennial, 1' tall, but may be prostrate
bearing 1–3 small pink flowers in
axil of most leaves along stem
with pinnate leaves. Flower
$1/16$" wide, petals none. Sepals
3, triangular. Stamens 3,
extend above sepals. Ovary
round inferior, 1 pistil, 3 pink
fringed stigmas. Fruit 3-
seeded, white hard drupe.
Stem lax, often in water.
Leaf types: (1) above-water leaf
linear-lanceolate, smooth, toothed,
pointed, alternate; (2)
submerged leaf, multipli-
pinnate, lobes hair-like, entire leaf
feathery. Habitat: swamps, marsh
shores. FL: July–August.

PURPLE LOOSESTRIFE

Lythrum salicaria

LOOSESTRIFE FAMILY   LYTHRACEAE

136

Perennial, 4' tall, bearing spike of reddish-purple
flowers with hairy tubular calyx at top
of leafy hairy stem with opposite
leaves. Flower ³/₄" wide, petals 6, long,
crinkled. Calyx 6-pointed, hairy,
shorter than petals; 6 short appendages alternate with petals. Stamens 12.
Ovary superior, 1 pistil, 1 style. Fruit capsule. Bract
below flower. Three flower types: (1) long stamens,
short style; (2) short stamens, long style; (3) short
and medium stamens, medium style. Leaves 3¹/₂"
long, entire, smooth, sessile. Habitat: wet meadows;
aggressive weed; introduced. FL: July–September.

FIREWEED

Epilobium angustifolium

EVENING-PRIMROSE FAMILY   ONAGRACEAE

Perennial, 6' tall, bearing spike of large
deep pink flowers at top of leafy stem.
Flower 1" wide, petals 4, spreading, oval
with claw. Sepals 4, linear, pointed.
Stamens 8, long. Ovary inferior, 1
pistil, finely hairy style, 4 curled
stigmas form cross. Fruit erect, 4-
angled, 3" long capsule; seeds
silk-tufted. Stalk includes long
inferior ovary subtended by bract.
Immature flowers pendant. Stalk of
opened flowers at right angle to stem.
Leaves 6" long, narrow ovate, entire,
smooth, short-petioled; alternate.
Habitat: burned woods, disturbed
areas. FL: July–September.

EASTERN WILLOW-HERB
Epilobium coloratum
EVENING-PRIMROSE FAMILY   ONAGRACEAE

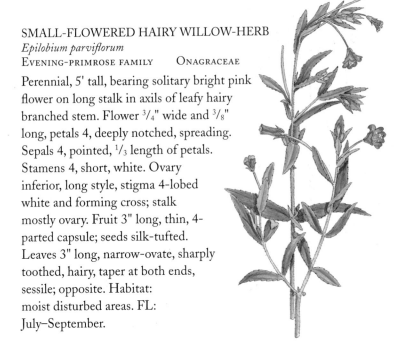

Perennial, 3' tall, bearing solitary pink flower on long dangling stalk in leaf axils of branched inflorescence on leafy stem. Flower $1/16$" wide, petals 4, notched. Sepals 4, red margin, hairy, tips protrude outwardly. Stamens 4, short. Ovary inferior, short style; most of stalk is ovary. Fruit 4-parted erect hairy capsule with red tip, seeds tufted with brown silk. Stem with incurved hairs, each branch with persistent large sessile leaf. Leaves 3" long, lanceolate, toothed, sessile; alternate above. Habitat: wet rocky shores. FL: July–September.

SMALL-FLOWERED HAIRY WILLOW-HERB
Epilobium parviflorum
EVENING-PRIMROSE FAMILY   ONAGRACEAE

Perennial, 5' tall, bearing solitary bright pink flower on long stalk in axils of leafy hairy branched stem. Flower $3/4$" wide and $3/8$" long, petals 4, deeply notched, spreading. Sepals 4, pointed, $1/3$ length of petals. Stamens 4, short, white. Ovary inferior, long style, stigma 4-lobed white and forming cross; stalk mostly ovary. Fruit 3" long, thin, 4-parted capsule; seeds silk-tufted. Leaves 3" long, narrow-ovate, sharply toothed, hairy, taper at both ends, sessile; opposite. Habitat: moist disturbed areas. FL: July–September.

FRINGED POLYGALA
Polygala paucifolia

MILKWORT FAMILY   POLYGALACEAE

Perennial, 6" tall, bearing 1–few
somewhat nodding irregular
pink-purple flowers at top of
leafy stem. Flower $3/4$" long,
corolla tube 3-lobed, 2 lobes
fused into upper hood, lower
lobe protrudes beyond hood
with fringed crest. Sepals 5,
irregular, 2 form large lateral wings,
others are small. Stamens 6, filaments fused,
under hood. Ovary superior, 1 pistil, 2-celled,
long style. Fruit small 2-seeded capsule.
Leaves $1^1/_2$" long, grouped at stem tip, oval,
smooth, pointed, petioled; alternate. Several
scale leaves on lower stem. Habitat: rich
sandy woods. FL: May–June.

NORTHERN WOOD-SORREL
Oxalis acetosella

WOOD SORREL FAMILY   OXALIDACEAE

Perennial, 6" tall, bearing solitary pink-striped flower on long leafless
stalk with basal trifoliate leaves. Flower
$3/4$" wide, petals 5, notched, marked
with darker pink veins. Sepals 5, small.
Stamens 10, short and long.
Ovary superior, 5-celled, 1
pistil, 5 styles. Fruit
elongated capsule.
Small bracts $3/4$ way
up flower stalk.
Rhizomatous stem
producing basal leaves
with long petiole; alternate.
Leaflets heart-shaped, entire,
fine downy hair. Habitat:
spruce-fir woods. FL: May–July.

WILD GERANIUM
Geranium maculatum
GERANIUM FAMILY   GERANIACEAE

Perennial, 2' tall, bearing cluster of
large fragrant pink-purple
flowers on hairy stalk at top of
hairy stem with lobed leaves.
Flower $1^1/_2$" wide, petals 5, $^3/_8$"
long, narrow to bearded claw.
Sepals 5, hairy, not glandular.
Stamens 5–10, above petals. Ovary
superior, 5-celled, persistent stigma.
Fruit erect capsule with long beak, appears like
crane's bill, brown pitted seeds. Basal leaf 3"
wide, palmately cleft into 5–7 elliptical lobes,
petioled. Stem with 1 pair sessile leaves,
palmately cleft into 5–7 elliptical lobes;
opposite. Habitat: moist rocky woods. FL: April–June.

HERB-ROBERT
Geranium robertianum
GERANIUM FAMILY   GERANIACEAE

Annual, 18" tall, bearing pairs of
pink flowers on long stalks in
upper axils of leafy hairy stem
with dissected leaves. Flower $^1/_4$"
wide, petals 5, with long claw,
often dark pink veins. Sepals 5,
awned, $^1/_2$ length of petals.
Stamens 5–10. Ovary superior, 1
pistil, long forked style. Fruit
long-beaked capsule, splits violently
to release brown wrinkled seeds.
Stem fragrant, hairy, branched, forms
runners. Leaves deeply 3–5 lobed, each
lobe pinnately dissected, upper leaflet
stalked. Habitat: borders of moist shaded
rocky woods. FL: May–September.

SPREADING DOGBANE
Apocynum androsaemifolium
Dogbane family   Apocynaceae

Perennial, 4' tall, bearing
small clusters of fragrant
nodding bell-shaped pink
flowers at top and in axils
on leafy stem with milky
sap. Flower $1/4$" wide,
corolla lobes pointed,
spreading, darker pink.
Calyx lobes short,
pointed. Stamens 5, short.
Ovary superior, 2-lobed, 5
basal nectaries, large stigma. Fruit 2 follicles,
nodding, seeds with tuft of hair. Leaves 3"
long, ovate, smooth, hairy underside, yellow
midrib, short petiole, drooping; opposite. Habitat: fields, borders of
woods. FL: June–August.

SWAMP-MILKWEED
Asclepias incarnata
Milkweed family   Asclepiadaceae

Perennial, 4' tall, bearing umbels of
pink flowers near top of leafy
stem with milky sap. Flower
$1/4$" wide, corolla 5 reflexed
red lobes. Calyx 5, under
petals. Stamens 5,
encircle style-stigma and together form
white-topped gynostegium at flower
center. Five cup-like hoods
about equal height
of gynostegium.
Five incurved pointed horns taller than
hoods; margin of hoods diverge, appear
scoop-like. Ovary superior, long. Seeds flat,
silk-tufted. Leaves 3" long, lanceolate, pointed, smooth, entire, sessile;
opposite. Habitat: swamps, wet ditches. FL: June–August.

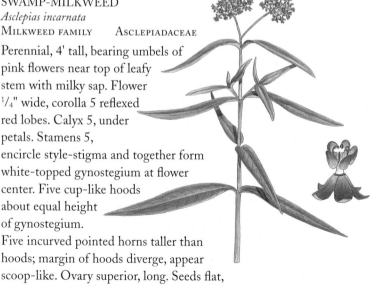

COMMON MILKWEED
Asclepias syriaca
MILKWEED FAMILY   ASCLEPIADACEAE

Perennial, 5' tall, bearing umbels of fragrant pink flowers on leafy stem with milky sap. Flowers $^1/_2$" wide, corolla 5 reflexed lobes. Calyx 5, small. Stamens 5, encircle style-stigma and together form gynostegium at flower center. Five cup-like hoods taller than gynostegium. Five incurved pointed horns shorter than hoods; extra lateral lobe with hood. Ovary superior, 2. Fruit warty pod; flat seeds with silky hairs. Leaves 10" long, broad, oval, pointed, entire, gray-downy underside, red midrib on old leaves; opposite. Habitat: old fields. FL: June–October.

HEDGE-BINDWEED
Calystegia sepium
MORNING-GLORY FAMILY   CONVOLVULACEAE

Perennial vine, 10' long, bearing large funnel-shaped pinkish flower with white stripes on stalk in axils of leafy vine with milky sap. Flower 3" wide, corolla 5 fused petals. Calyx 5-lobed, subtended by 2 heart-shaped bracts. Stamens 5, encircle style. Ovary superior, 1 pistil, 1 style. Fruit capsule. Stem encircles host for support. Leaves 4" long, triangular, arrow-shaped with basal lobes, smooth, entire, long-petioled; alternate. Habitat: thickets, moist disturbed areas. FL: May–September. Field-b., *Convolvulus arvensis,* similar but smaller flowers, small bracts at calyx base. Habitat: roadsides, dry disturbed areas; introduced.

LOPSEED
Phyrma leptostachya

VERVAIN FAMILY   VERBENACEAE

Perennial, 3' tall, bearing
spikes of opposite irregular
pink flowers at top and in
axils of leafy smooth stem.
Flower $^3/_8$" long, corolla pink
and white, 2-lipped. Pink
lower lip short, 2–3 lobed.
White upper lip long, broad,
3-lobed, these tending to
infold. Calyx tube 2-lipped, the
large lip with 3 dark-pink teeth. Stamens 4. Ovary superior, 1 pistil, long forked
style. Fruit achene, persistent calyx pendant
and appressed to flower spike. Leaves 5" long, ovate, pointed, toothed,
smooth, petioled; opposite. Habitat: moist woods. FL: June–August.

MOTHERWORT
Leonurus cardiaca

MINT FAMILY   LAMIACEAE

Perennial, 4' tall, bearing clusters of fragrant irregular pink flowers in
axils on leafy smooth square stem with forked
leaves. Flower $^1/_4$" long, corolla 2-lipped,
bearded, upper lip arched over 3-lobed
lower lip with red-violet spots.
Calyx tube sharp-spined,
lower 2 spines recurved;
persistent. Stamens 4, under
hood. Ovary superior, 4-lobed,
1 style. Fruit smooth
nutlets. Lower leaves 4"
long, center fork longest,
coarsely toothed, petioled,
veins with bumpy surface. Upper leaves
3-lobed, toothed, base narrows to wedge,
petioled; opposite. Habitat: disturbed areas;
introduced. FL: June–August.

WILD BASIL
Satureja vulgaris
MINT FAMILY   LAMIACEAE

Perennial, 18" tall, bearing many irregular violet flowers
in terminal domed clusters on leafy hairy square
stem. Flower $^1/_2$" long, corolla tube 2-lipped, upper
lip notched, lower lip 3-lobed. Calyx tube 2-lipped,
pointed lips, hairy. Stamens 4, under upper lip.
Ovary superior, 1 pistil, short forked style. Fruit 4
smooth nutlets. Hairy bracts below each flower;
they equal length of calyx and impart woolly
appearance to head. Lower leaves 3" long, lan-
ceolate-oval, smooth to toothed, and upper leaves
$1^1/_2$" long, elliptical; opposite. Habitat: wood edges;
introduced. FL: June–September.

AMERICAN GERMANDER
Teucrium canadense
MINT FAMILY   LAMIACEAE

Perennial, 2' tall, bearing terminal
spikes of irregular pink-violet
flowers on leafy downy square
stem. Flower $^1/_2$" long, corolla 2-
lipped. Upper lip short with 2
tiny lobes. Lower larger lip with
wide flat middle lobe and 2
small lateral lobes. Calyx 5-
toothed, short. Stamens 4,
arched, projecting from corolla.
Ovary superior, 4-lobed, style
length of stamens. Fruit 4 nearly
united nutlets. Stem short-
branched near top. Leaves 4" long,
lanceolate, sessile, toothed, hairy
underside; opposite or alternate. Habitat:
wood thickets, shores. FL: June–September.

SMOOTH AGALINIS
Agalinis purpurea
<small>FIGWORT FAMILY</small>   <small>SCROPHULARIACEAE</small>

Annual, 3' tall, bearing bright pink irregular
bell-shaped flower in axils near top of leafy
smooth squarish stem. Flower $1/2$" wide,
corolla 5 spreading lobes, 2-lipped, downy
inside. Calyx 5-lobed, pointed, $1/4$ length of
corolla. Stamens 4, yellow anthers. Ovary
superior, 1 pistil, 1 style. Fruit globose
capsule. Flower stalk shorter than subtending leaf and about length of calyx. Leaves
$1^1/2$" long, narrow, pointed, entire, dark
edges, sessile; opposite. Habitat: moist
meadows and shores. FL: July–September.
Common A., *A. tenuifolia*, stalk longer than
subtending leaf and calyx. Habitat: moist
meadows.

TWINFLOWER
Linnaea borealis
<small>HONEYSUCKLE FAMILY</small>   <small>CAPRIFOLIACEAE</small>

Perennial trailing vine, 4" tall, bearing pairs of fragrant nodding
funnel-shaped pink flowers at top of leafy stem. Flower $1/4$" wide,
corolla 5-lobed, dark pink, hairy inside. Calyx 5-lobed, small. Stamens
4, within corolla. Ovary inferior, slender style,
globose stigma. Fruit dry
1-seeded berry. Flower
has 2 glandular bracts
beneath. Stem prostrate,
dark red, forms erect
leafy flower stems at
nodes. Leaves $3/4$"
long, round, upper
half toothed,
smooth, short-
petioled; opposite.
Habitat: wet woods, bogs.
FL: June–August.

SNOWBERRY
Symphoricarpos albus
HONEYSUCKLE FAMILY   CAPRIFOLIACEAE

Perennial, 3' tall, bearing pairs of nodding bell-shaped pink flowers in axils of top leaves on leafy stem. Flower $^1/_4$" wide and long, corolla 5 spreading lobes, hairy inside. Calyx 5-pointed, very small. Stamens 5, inside corolla. Ovary inferior, 1 pistil, stigma-style to top of corolla. Fruit waxy white 2-seeded berry. Leaves 2" long, oval, entire, gray-green, short petiole; opposite. Habitat: rocky woods, roadsides. FL. May–July.

PHILADELPHIA DAISY
Erigeron philadelphicus
DAISY FAMILY   ASTERACEAE

Perennial, $3^1/_2$' tall, bearing several flower-heads with many feather-like pink rays and yellow center at top and in axils of leafy hairy stem. Head 1" wide, over 150 ray flowers. Disk flower tubular, short. Calyx pappus of long bristles. Ovary inferior, 1 pistil, style lobes appendaged. Fruit 2-nerved achene, pappus. Involucral bracts in 1 ring, short, hairy. Basal leaves 6" long, elliptical, wavy-toothed, round tip, hairy underside. Stem leaves shorter, base clasps stem, toothed, hairy underside, velvety. Habitat: disturbed areas. FL: May–September.

MIST-FLOWER
Eupatorium coelestinum
DAISY FAMILY   ASTERACEAE

Perennial, 3' tall, bearing flat-topped cluster of dusty pink flower-heads on leafy hairy stem. Head $^3/_{16}$" wide, 35–70 disk flowers. Corolla long whitish tube with violet teeth. Calyx pappus of white bristles. Ovary inferior, very long forked white styles extend above corolla. Young head of upright disk flowers with long styles surrounded by pink involucral bracts (left). Mature head of spreading corollas surrounded by spreading violet involucral bracts (right). Fruit smooth 5-angled achene, pappus. Stem dotted red, fine hairs. Leaves 4" long, ovate, short petiole, toothed, finely hairy underside; opposite. Habitat: moist meadows, shores. FL: July–October.

SPOTTED JOE-PYE WEED
Eupatorium maculatum
DAISY FAMILY   ASTERACEAE

Perennial, 5' tall, bearing flat-topped clusters of small pink flower-heads at top of leafy stem with whorled leaves. Heads $^1/_4$" wide, 8 or more disk flowers only. Corolla tubular, 5 spreading lobes. Calyx pappus of fine bristles. Stamens 5, encircle style. Ovary inferior, 1 pistil, very long branched style. Fruit black 5-angled glandular achene, pappus. Involucral bracts in several rings, purplish, 3-nerved. Stem red-purple. Leaves 8" long, lanceolate, pointed, thick, toothed, short petiole; 3–5 whorled. Habitat: moist meadows, alkaline soils. FL: July–September.

TWISTED STALK
Streptopus roseus
LILY FAMILY   LILIACEAE

Perennial, 3' tall, bearing
single small nodding bell-
shaped pink flower on
twisted stalks in axils on
forked zigzag leafy stem.
Flower $^3/_8$" long, 6 tepals
with dark pink spots inside.
Stamens 6, double-pointed,
within tube. Ovary superior, short
style, 3-lobed stigma. Fruit $^3/_8$" red
berry on dangling stalk. Stem arching,
forked at mid-height, clasping leaves
below fork. Leaves 6" long, lanceolate,
sessile, pointed, some leaves clasp stem;
alternate. Habitat: moist rich woods. FL: April–July.

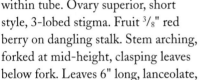

DRAGON'S MOUTH
Arethusa bulbosa
ORCHID FAMILY   ORCHIDACEAE

Perennial, 10" tall, bearing showy
irregular pink flower with flat lip on
green leafless stalk. Flower 1" wide,
petals 3, lower lip flat, broad, pendant,
yellow-crested and pink-spotted,
fringed toward tip; 2 lateral petals
narrow-spatulate, pointed. Sepals 3,
narrow-spatulate, pointed, pink. Stamen
anther fused with style to form winged
column which projects over crested lip.
Ovary inferior, erect. Fruit capsule. Flower
bract scale-like triangle. Stalk with several
sheathing scales. Leaf 6" long, lanceolate,
appears after flower. Habitat: bogs near
open pool. FL: May–August.

GRASS-PINK
Calopogon tuberosus
ORCHID FAMILY   ORCHIDACEAE

Perennial, 1' tall, bearing several fragrant
irregular rose-purple flowers on leafless stalk
with one grass-like basal leaf. Flower orienta-
tion reversed that of other orchids. Flower 1"
wide, petals 3, upper petal (lip) spatulate with 3
rows of fleshy yellow-tipped hairs; lower petals
spreading. Sepals 3, oval-lanceolate, lower one
curved up; upper sepals spreading. Stamen anthers
2, fused with style to form long column with rose-
purple spoon-like tip. Ovary inferior, 1 pistil, 1
style. Fruit capsule. Leaf $^1/_2$ height of plant.
Habitat: wet meadows, bogs. FL: June–August.

CALYPSO
Calypso bulbosa
ORCHID FAMILY   ORCHIDACEAE

Perennial, 8" tall, bearing showy
irregular pink flower with slipper-
like lip on leafless stalk above basal
leaf. Flower $^3/_4$" wide, petals 3, lip
petal inflated white and pink-striped
slipper with 2 projections at tip and
bearded with 3 rows of yellow hairs;
other 2 petals and 3 sepals narrow,
pink, spreading above lip.
Stamen anther fused with
style to form hooded column
over lip. Ovary inferior. Fruit
capsule. Pink bracts below
flower. Stalk pink with 2–3 scales.
Leaf 3" long, oval, blunt-tipped,
petioled. Habitat: wet mossy woods.
Threatened. FL: May–June.

STRIPED CORAL-ROOT
Corallorhiza striata
ORCHID FAMILY   ORCHIDACEAE

Perennial, 2' tall, bearing several irregular reddish-pink flowers with dark maroon lips on leafless stem. Flower $1/2$" long, 3 petals, lower lip petal tongue-like, dark maroon, with 2 additional short pink stripes near its base; 2 lateral pink-striped petals incurved. Three pink-striped sepals curve toward lip, upper one forms hood over lip. Stamen anthers 2, fused with style to form tip of column above stigma; lid covers anther sacs. Ovary inferior. Fruit capsule. Stem red-purple, darker base. Light purple scales on stem. Habitat: rich woods. FL: May–July.

MOCCASIN-FLOWER
Cypripedium acaule
ORCHID FAMILY   ORCHIDACEAE

Perennial, 16" tall, bearing one irregular flower with pink slipper pouch on leafless hairy stem with basal leaves. Flower 2" long, petals 3, lower lip petal inflated into pink-veined slipper with narrow central opening; lateral petals narrow, twisted, inner basal area hairy. Sepals 3, upper sepal arches over slipper; laterals fused into lower sepal. Stamen anthers 2, fused with style to form column projecting toward pouch; staminode covers stigma. Ovary inferior. Fruit capsule. Bract below flower. Two basal leaves 8" long, elliptical, glandular. Habitat: sandy woods. FL: May–July.

ROSE POGONIA
Pogonia ophioglossoides
ORCHID FAMILY   ORCHIDACEAE

Perennial, 18" tall, bearing showy irregular pink flower with long lip on stem with one leaf. Flower 1½" wide, petals 3, lip petal long, fringed margin, center bearded with yellow hairs; lateral 2 dark-pink petals oval, arching over lip. Sepals 3, pink, spatulate, dorsal sepal erect, lateral sepals spreading. Stamen anther fused with style to form winged column that projects over lip. Ovary inferior, 1-celled, erect. Fruit capsule. Flower with leaf-like bract below. Leaf 1½" long, narrow-oval, smooth. Habitat: wet open bogs. FL: May–August.

Green

FLOWERS

EARLY MEADOW-RUE
Thalictrum dioicum
BUTTERCUP FAMILY   RANUNCULACEAE

Perennial, 30" tall, bearing long terminal inflorescence stalk with many drooping greenish-maroon flowers on stem with compound leaves. Separate male and female plants. Male: petals none, sepals 4, many maroon stamens with yellow-green anthers (shown). Female: petals none, sepals 4, ovaries superior, several elongated purplish pistils with short pointed stigmas, fruit ribbed achene with curved persistent stigma. Leaves long-petioled, 3× divided, each round 1" leaflet 3-lobed and each lobe toothed toward tip, pale yellow underside. Habitat: moist rich woods. FL: April–May.

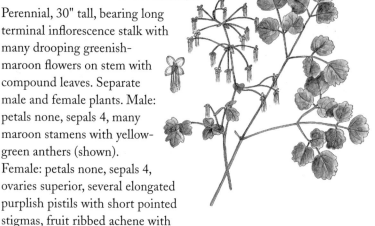

BLUE COHOSH
Caulophyllum thalictroides
BARBERRY FAMILY   BERBERIDACEAE

Perennial, 3' tall, bearing cluster of green-yellow flowers at top of stalk on smooth stem with 3-parted leaves. Flower $^3/_8$" wide, 6 tiny green-yellow petals, cup-like, modified into glands. Large sepals 6, green-yellow with brown stripes, pointed, petal gland at base. Bracts 4, green, below sepals. Stamens 6, yellow spots under hood. Ovary superior, 1 pistil, short lateral stigma. Fruit $^1/_4$" blue berry. Leaves large, 2× trifoliate into leaflets 2" long and each leaflet 3-lobed, tip of large leaflet often 3-pointed. Habitat: moist, rich woods. FL: April–May.

COMMON HOPS
Humulus lupulus
INDIAN HEMP FAMILY   CANNABACEAE

Perennial vine, 10' long, bearing small cluster of green flowers in axils on coarsely barbed glandular stem with lobed leaves. Plants of male or female flowers. Female $1/16$" wide, petals none, sepals 5, superior ovary, elongated style, fruit an achene; small bract below each pair of flowers. As fruits mature bracts enlarge to surround fruit forming catkin (shown right). Male $1/16$" wide, sepals 5, stamens 5, small bract below each pair of flowers; flowers in axillary spike. Leaves 5" wide, round, 3–5 lobed, heart-shaped base, toothed, petioled; opposite. Habitat: edge of maple-beech woods. FL: July–August.

WOOD NETTLE
Laportea canadensis
NETTLE FAMILY   URTICACEAE

Perennial, 4' tall, bearing dangling clusters of tiny green flowers in axils of leafy hairy stem with stinging hairs. Separate male and female flowers, no petals. Male: short clusters in lower nodes, sepals 5, stamens 5, white. Female: long clusters in upper nodes, sepals 4 the outer 2 smaller, superior ovary, long hairy style. Fruit flat, winged, black crescent achene. Leaves 8" long, ovate, toothed, pointed, sharp tip, thin, petioled; alternate. Habitat: moist woods, streambanks. FL: July–September.

STINGING NETTLE
Urtica dioica, var. *procera*
NETTLE FAMILY   URTICACEAE

Perennial, 7' tall, bearing dangling
clusters of tiny green flowers in
upper axils of leafy 4-angled stem
with stinging hairs. Male and
female flowers on same or different
plants. Female in lower axils:
petals none, sepals 4, finely hairy
superior ovary with tufted stigma
and 1 large stinging hair (above).
Male in upper axils: petals none,
sepals 4, stamens 4, nonfunc-
tional ovary (below). Fruit
achene enclosed by sepals. Leaves 6" long,
narrow-ovate, stinging hairs on veins, toothed, petioled, linear stipules;
opposite. Habitat: moist creekbeds. FL: June–September.
Variety *dioica* similar but with numerous stinging hairs. Habitat: moist
creekbeds.

LAMB'S QUARTERS
Chenopodium album
GOOSEFOOT FAMILY   CHENOPODIACEAE

Annual, 3' tall, bearing terminal and lateral
spikes of tiny green flowers on leafy
smooth branched stem. Flower $1/16$" wide,
petals none. Calyx 1–5, small lobes.
Stamens 1–5. Ovary superior, 1 pistil, 2
styles. Fruit dry 1-seeded utricle, shiny
black seed, surface sculptured. Flower
spikes become reddish late in season.
Lower leaves 3" long, triangular, coarsely
toothed, gray underside, petioled; alternate. Upper leaves among spikes small,
lanceolate, smooth, entire, petioled. Habitat: disturbed areas,
fields, roadsides; introduced.
FL: June–October.

OAK–LEAVED GOOSEFOOT
Chenopodium glaucum
GOOSEFOOT FAMILY   CHENOPODIACEAE

Annual, 20" tall, bearing small clusters of tiny green flowers in axils of usually procumbent succulent leafy stem. Flower $1/16$" wide, petals none. Calyx 4 or 5 small green lobes joined at base, red-tipped, free from seed. Stamens 2–4. Ovary superior, broad stigma. Fruit dry 1-seeded utricle, shiny dark brown flat seed, surface roughened. Stem smooth, reddish on one side and green on other, branched from base. Lower leaves elliptical, lobed, mealy-white underside, petioled. Upper leaves lobed, narrow, often sessile; alternate. Habitat: moist sandy shores, introduced. FL: June–September.

ROUGH PIGWEED
Amaranthus retroflexus
AMARANTH FAMILY   AMARANTHACEAE

Annual, 3" tall, bearing clusters of tiny green flowers and bracts on branched rough leafy stem. Clusters of male or female flowers. Male: petals none; sepals 5; stamens 5, subtended by rigid bracts longer than sepals. Female: petals none; sepals 5, lanceolate; superior oval ovary, 1 style, subtended by 3 rigid bracts longer than ovary. Fruit sparkling inflated bright pink utricle around brown seeds (shown); utricle shorter than bracts. Stem minutely hairy. Leaves 4" long, elliptical, petioled, entire; alternate. Habitat: disturbed areas; introduced. FL: July–October.

WILD CUCUMBER
Echinocystis lobata
GOURD FAMILY   CUCURBITACEAE

Annual vine, 15' long, bearing
separate stalked greenish male
and female flowers in axils of
leafy vine with tendrils.
Flowers female (above)
or male (below). Male: in
clusters, corolla 6 long petal-like
lobes, calyx 6 short pointed lobes,
stamens 3, short, joined. Female:
solitary, often in axil with male,
corolla 6 petal-like lobes on top spiny inferior
ovary, calyx 6 short pointed lobes, short stigma.
Fruit inflated spiny berry. Stem angular, grooved.
Leaves 3" long, rough, 3–7 lobes; alternate. Habitat: moist ground,
thickets. FL: July–September.

GREENISH-FLOWERED SHINLEAF
Pyrola chlorantha
SHINLEAF FAMILY   PYROLACEAE

Perennial, 10" tall, bearing several
waxy fragrant nodding greenish-
white flowers striped green on
smooth stalk with basal leaves.
Flower $^1/_4$" wide, petals 5, oblong.
Sepals 5, short. Stamens 10, shorter
than petals. Ovary superior, curved
greenish-yellow stigma-style longer
than petals. Fruit capsule. Basal leaves
$^5/_8$" wide, oval, blade $1^1/_4$" long, entire,
blade $^1/_2$ length of petiole, distinct
venation. Habitat: sandy
woods. FL: June–July.
Elliptic S., *P. elliptica*,
similar but blades longer,
triangular sepals, 10 or more
flowers. Habitat: upland woods.

NAKED MITREWORT
Mitella nuda
SAXIFRAGE FAMILY   SAXIFRAGACEAE

Perennial, 8" tall, bearing short spike of several
very delicate greenish-white flowers
on leafless downy flower stalk
adjacent to basal leaves. Flower $^3/_{16}$"
wide, petals 5, thread-like, fringed,
snowflake appearance. Sepals 5,
green, oval, pointed. Stamens 10, short,
white anthers. Ovary partly inferior, 1
pistil, 2 short styles. Fruit 2-beaked
mitre-like capsule, many seeds. Flower
stalk pink, downy. Leaves $^3/_4$" wide, hairy,
round-toothed, heart-shaped, long-petioled.
Habitat: wet woods often on moss.
FL: May–June.

LADY'S MANTLE
Alchemilla vulgaris
ROSE FAMILY   ROSACEAE

Perennial, 2' tall, bearing small cluster
of tiny greenish-yellow flowers on
hairy leafy stem with palmately
lobed leaves. Flower $^3/_{16}$" wide,
petals none. Sepals 4, yellow-
green, spreading, triangular.
Bracts 4, green, below
sepals. Stamens 4, black
anthers arranged as square.
Ovary small, cup-like with
narrow orange throat, 1
pistil, short style. Fruit
achene. Leaves 4" wide,
velvet hairy, round, multi-
lobed and toothed, overlapping leaf
base, large stipule; lower leaves long-
petioled, upper sessile. Habitat:
roadsides; introduced. FL: May–August.

COOPER'S MILK-VETCH

Astragalus neglectus

PEA FAMILY   FABACEAE

Perennial, 2' tall, bearing clusters of
irregular greenish flowers on leafy
smooth stem. Flower $5/8$" long,
corolla 2-lipped, upper lip
spatulate with lobed up-
turned tip, lower lip a keel
of 2 fused broad lobes, 2
lateral elongated lobes incurved against keel. Calyx
tubular $1/3$ length corolla, 5-
toothed, minutely hairy. Stamens 10. Ovary
superior, 1 pistil. Fruit smooth oval nearly sessile sharp-pointed pod, 1
chamber. Leaves pinnate, small pointed stipule; alternate. Leaflets $3/4$"
long, 6–12 pairs, linear, entire, minutely hairy underside, fold together
at night. Habitat: moist calcareous soil. *Endangered.* FL: July.

LEAFY SPURGE

Euphorbia esula

SPURGE FAMILY   EUPHORBIACEAE

Perennial, 3' tall, bearing clusters
of yellow-green flowers on long
stalks subtended by whorl of 7–8
leaves on smooth stem with
milky sap. Male and female
flowers in cup-shaped brown
cyathium. Green-brown
crescent glands, 4, line edge of cyathium
(shown). Yellow-green heart-shaped
bracts, 2, below cyathium; 2 green buds
adjacent to cyathium. In cyathium: male
is 1 stamen, several present; female is 1 ovary
with 3 stigmas. Fruit capsule. Maturing ovary
grows out of cyathium on curved stalk. Leaves 2" long × $1/4$" wide,
lanceolate, sessile. Habitat: pastures; introduced. FL: June–August.
Cypress-s., *E. cyparissias*, similar but shorter, narrow $1/8$" wide leaves.
Habitat: roadsides; introduced.

SEASIDE-SPURGE
Euphorbia polygonifolia
SPURGE FAMILY   EUPHORBIACEAE

Creeping annual, 8" long, bearing green flowers in axils of branched leafy smooth succulent stem with milky sap. Male and female flowers in cupped cyathium subtended by 2 bracts. Cyathium edge with 4–5 small reflexed greenish lobes alternating with small reddish glands. In cyathium, male is 1 stamen, several; female is 1 ovary with 3 stigmas (above). Fruit capsule. Ovary stalk grows out of cyathium during maturation (below). Leaves $1/2$" long, narrow, sessile; opposite. Habitat: sandy shore. *Special Concern.* FL: July–September. Spotted-s., *E. maculata,* similar but leaves have reddish spot. Habitat: lawns, disturbed areas.

FROST-GRAPE
Vitis riparia
GRAPE FAMILY   VITACEAE

Perennial vine bearing cluster of fragrant green flowers on stalk opposite leaf near base of young shoot with tendrils. Separate male and female flowers, $1/16$" wide, calyx none. Corolla 5-lobed, joined at tips and detached at base. Male: 5 stamens, long filaments, with detached corolla (shown). Female: ovary with style, aborted stamens. Fruit black smooth berry. Leaves 5" wide, round, shallow pointed lobes, smooth, coarsely toothed, petioled; opposite, often paired with tendril. Young leaf underside hairy, but only on veins when mature. Habitat: woods, thickets. FL: June.

POISON-IVY
Toxicodendron radicans
CASHEW FAMILY   ANACARDIACEAE

Perennial, 2' tall or vine, bearing many small green flowers in clusters on stalks in axils of trifoliate leaves on leafy stem. Do not touch! Flower $1/8$" wide, petals 5, oval, spreading. Sepals 5, small, pointed. Stamens 5, projecting above petals. Ovary superior, 1-celled, 1 style. Fruit whitish drupe. Stem hairy, produces aerial roots to clasp host. Leaflets 4" long, ovate, often occur as 3 glossy asymmetrical lobed leaflets, upper leaflet long-petioled, red-tinted upon aging. Plant oils cause skin rashes. Habitat: disturbed areas, roadsides. FL: May–July.

WILD SARSAPARILLA
Aralia nudicaulis
GINSENG FAMILY   ARALIACEAE

Perennial, 18" tall, bearing 3 umbels of many tiny green flowers on leafless stalk overtopped by adjacent pinnate leaf. Flower $1/16$" wide, petals 5, reflexed. Sepals none. Stamens 5, white anthers on long filaments. Ovary inferior, 5-celled, short style. Fruit black berry. Bracts at base of umbel. Stem rhizome produces separate flower stalk and long-petioled leaf taller than stalk. Petioled leaf 3-parted, each part pinnately divided into 3 or 5 leaflets. Leaflets elliptical, pointed, toothed, smooth, petioled, base of laterals may be asymmetric. Habitat: rich upland woods. FL: July–August.

CLUSTER-SANICLE
Sanicula gregaria
CARROT FAMILY   APIACEAE

Perennial, 2' tall, bearing small umbels of green flowers at top of stem with trifoliate leaves. Separate male and complete flowers. Complete flower $1/16$" wide, corolla 5-lobed, cup-shaped. Calyx 5-lobed, reddish-hairy. Stamens 5. Ovary inferior, 2 short styles. Fruit 2-seeded, round with hooked bristles, style shorter than bristles. Lower leaves long-petioled, palmately 3-parted but appear 5 because lateral leaflets are deeply cleft; leaflets 2" long, toothed, lobed. Upper leaves similar to lower but smaller, sessile. Habitat: moist woods. FL: June–August.

BLACK SNAKEROOT
Sanicula marilandica
CARROT FAMILY   APIACEAE

Perennial, 2' tall, bearing small umbels of green flowers on leafy stem with palmate leaves. Separate male and female flowers often present. Flower $1/16$" wide, 5 green-yellow petals, short, bases fused. Calyx 5, short points. Stamens 5, short. Ovary inferior, 2 long styles. Fruit narrow seed with hooked bristles, style longer than bristles. Lower leaves long-petioled, palmately 5-parted, but appear like 7 because lateral pair deeply lobed; leaflets 3" long, elliptical, toothed. Stem leaves lanceolate, toothed, sessile, 3-parted, lower leaflets deeply lobed; opposite. Habitat: moist woods. FL: June–August.

SPURRED-GENTIAN
Halenia deflexa
GENTIAN FAMILY   GENTIANACEAE

Annual, 6" tall, bearing several
spurred nodding green-yellow
flowers at top of leafy stem. Flower
³/₈" long, corolla with narrow throat,
4 long spurs. Dark green calyx 4-
pointed, each alternating with spur.
Stamens 4, within corolla. Ovary
superior, 1 stigma-style. Fruit
capsule. Leaves 1¹/₂" long, ovate,
blunt-tipped, smooth, entire,
sessile, veins prominent; opposite.
Habitat: bogs, moist evergreen
woods. FL: July–August.

WILD LICORICE
Galium lanceolatum
MADDER FAMILY   RUBIACEAE

Perennial, 1' tall, bearing
small greenish-white
flowers in loose terminal
cluster at top of leafy
smooth stem with whorled
leaves. Flower ³/₁₆" wide,
corolla 4 oval pointed lobes,
smooth. Sepals none.
Stamens 4, brown anthers.
Ovary inferior, 2 short styles,
globose stigmas. Older
flowers purple. Fruit 2-lobed
dry capsule with hooked hairs
that stick to fur, clothing. Leaves 2"
long, ovate, sessile or petioled; whorls of 4.
Habitat: dry woods, thickets. FL: June–July.

SWEET-SCENTED BEDSTRAW
Galium triflorum

MADDER FAMILY   RUBIACEAE

Reclining perennial, 3' long, bearing 3 fragrant tiny green flowers on long stalk in axils near top of leafy stem with whorled leaves. Flower ⅛" wide, corolla 4 incurved tips. Sepals none. Stamens 4. Ovary inferior, short style, globose stigma. Fruit 2-lobed dry capsule, hooked hairs stick to fur, clothing. Stem smooth or few hairs, lax, forming mats. Leaves 1" long, lanceolate, sessile, bristle-tipped, fine stiff hairs on margin point toward tip; whorls of 6. Habitat: moist woods. FL: June–August.

COMMON RAGWEED
Ambrosia artemisiifolia
DAISY FAMILY   ASTERACEAE

Annual, 4' tall, bearing spikes of green flower-heads in axils of finely hairy stem with dissected leaves. Flower ¹⁄₁₆" wide, male near top, female in axils. Male: cupped receptacle with 5–12 points containing 6 yellow anthers. Female: inferior ovary, 1 pistil, short style; no corolla or calyx. Involucral bracts nearly closed, encircle achenes tipped with 6 spines. Leaves 4" long, 2× divided into narrow leaflets; alternate. Habitat: disturbed areas; alien, source of hayfever pollen. FL: July–October. Giant R., *A. trifida*, similar but taller, 3-lobed pointed leaves

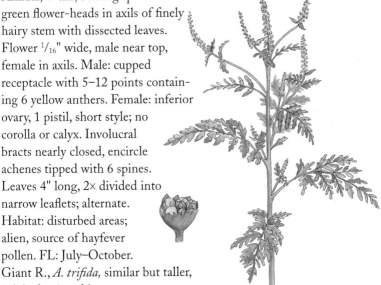

WORMWOOD
Artemisia campestris
<small>DAISY FAMILY ASTERACEAE</small>

Biennial, 3' tall, bearing small green flower-heads in loose spike on smooth stem with dissected leaves. Head 1/4" wide, disk flowers only. Outer ring of 6–8 yellow flowers. Inner flowers sterile, but each with long style. Corolla 5 pointed lobes. Calyx pappus none. Ovary inferior, 1 pistil, flat style, fringed stigma. Receptacle without hairs. Fruit smooth achene. Involucral bracts green-yellow, broad, scaly, curved tips; outer, green. Leaves 4" long, 3× divided into thread-like leaflets; upper smaller. Habitat: sandy areas. FL: July–August.

COSTMARY
Chrysanthemum balsamita
<small>DAISY FAMILY ASTERACEAE</small>

Perennial, 4' tall, bearing clusters of fragrant green flower-heads on branches of leafy stem. Heads 1/4" wide, many disk flowers encircled by white receptacle ring. Corolla 5-lobed. Calyx pappus absent. Stamens 5, encircle style. Ovary inferior, 1 pistil, style short, forked, flat with short hairs. Fruit angular ribbed achene, forked beak. Involucral bracts narrow with expanded tip. Basal leaves 8" long, thickened, elliptical, toothed. Stem leaves smaller, gray-green, elliptical, small basal lobes clasp stem, sessile, toothed; alternate. Habitat: moist meadows; introduced. FL: August–September.

JACK-IN-THE-PULPIT
Arisaema triphyllum
ARUM FAMILY   ARACEAE

Perennial, 2' tall, bearing greenish-pink
flower stalk with spathe or Pulpit over-
arching pink spadix or Jack
and adjacent leaves. Spathe
interior streaked, mottled
purple. Separate male and
female flowers, or plants.
Female: many near spadix
base, ovary with stigma. Male: above female,
2–5 stamens. Fruit $1/4$" red berries on
spadix. Stem, corm, forms separate leaf and
flower stalks. Leaves 2' tall, long petiole,
trifoliate blade, leaflets 6" long, elliptical,
sessile, smooth, over-top stalk. Spadix odoriferous.
Habitat: moist to damp woods, swamps. FL: April–July.

BUR-REED
Sparganium americanum
BUR-REED FAMILY   SPARGANIACEAE

Perennial aquatic, 3' tall, bearing
globose heads of many green
flowers on zigzag stem. Upper
small heads male and lower wider
heads female flowers, in axil of long
keeled bract. Flower-head axis
unbranched. Male 3–6 scaly tepals,
stamens 5. Female 3–6 scaly tepals,
ovary superior, 1 pistil with 1 stigma;
head of many pistils (mature
female shown). Fruit achene.
Leaves 3' long, flat, smooth,
keeled at leaf base; alternate. Habitat:
shallow water, shores. FL: May–August.
Giant B., *S. eurycarpum*, 2 stigmas,
flower-head axis branched. Habitat:
shallow water, shores.

SOLOMON'S SEAL
Polygonatum pubescens
LILY FAMILY   LILIACEAE

Perennial, 3' tall, bearing 1–2 pendant bell-shaped green flowers from leaf axils on arching leafy stem. Flower $1/2$" long, perianth long tube, 6 small tepal lobes at mouth, 3 inner lobes light green, 3 outer lobes dark green with midvein. Stamens 6, half height of perianth. Ovary superior, 3-celled, globose stigma. Fruit several seeded blue berry. Stem broadly arching. Leaves 5" long, elliptical, pointed, finely hairy underside but smooth above, short-petioled, entire, parallel veins; alternate. Leaf scar on rhizome appears like seal. Habitat: moist to medium woods. FL: May–June.

BRISTLY GREENBRIER
Smilax hispida
CATBRIER FAMILY   SMILACACEAE

Perennial vine, 10' long, bearing umbels of green flowers in axil of leafy prickly stem with tendrils. Separate male (left) and female flowers, $1/2$" wide. Male: perianth 6 deeply cleft tepals, stamens 6. Female: perianth 6 deeply cleft reflexed tepals, ovary superior, 3 styles. Fruit blue berry. Umbel stalk longer than leaf petiole. Stem spiny near base (below), fewer to no spines above, tendril grows from each tip of stipule. Leaves 4" long, ovate, pointed, round base, petioled; alternate. Habitat: moist thickets. FL: May–June.

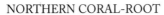

NORTHERN CORAL-ROOT
Corallorhiza trifida
ORCHID FAMILY   ORCHIDACEAE

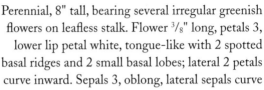

Perennial, 8" tall, bearing several irregular greenish flowers on leafless stalk. Flower $^3/_8$" long, petals 3, lower lip petal white, tongue-like with 2 spotted basal ridges and 2 small basal lobes; lateral 2 petals curve inward. Sepals 3, oblong, lateral sepals curve inward toward lip, upper sepal erect forming hood over petals and lip. Stamen anthers 2, fused with style to form column, its tip above stigma. Ovary inferior, yellowish, nodding. Fruit elongated capsule, many seeds. Stem smooth, green. Scales on stem. Habitat: maple-beech woods. FL: May–June.

HOOKER'S ORCHID
Habenaria hookeri
ORCHID FAMILY   ORCHIDACEAE

Perennial, 2' tall, bearing several irregular bractless green flowers on leafless stem with 2 large basal leaves. Flower $^3/_8$" long, petals 3, lower lip petal lanceolate, its base turned backward into long brown spur; upper petals elongate and curve inward. Sepals 3, oval, upper sepal arched as hood over column; lower 2 spreading. Stamen anthers 2, fused with style to form column above opening of spur. Ovary inferior. Fruit capsule. Leaves 6" wide, round. Habitat: hillsides of rich woods. *Special Concern.* FL: June–July.

TALL NORTHERN BOG-ORCHID
Habenaria hyperborea
ORCHID FAMILY   ORCHIDACEAE

Perennial, 2' tall, bearing elongated cluster of irregular green flowers at top of leafy stem. Flower $^3/_4$" long, petals 3, lower lip petal linear, extending back into curved spur as long as lip; upper yellow petals lanceolate, arched inward. Sepals 3, oblong, upper sepal arched as hood above column, lateral 2 sepals spreading. Stamen anthers 2, fused with style to form column with stigma tip at opening of spur. Ovary inferior. Fruit capsule. Leaves 6" long, narrow, sheath stem; upper become shorter. Bract below flower; alternate. Habitat: wet woods, bogs. FL: June–July.

BLUNT-LEAVED ORCHID
Habenaria obtusata
ORCHID FAMILY   ORCHIDACEAE

Perennial, 6" tall, bearing several irregular greenish-white flowers on leafless stalk with basal leaf. Flower $^3/_{16}$" long, petals 3, lip petal strap-shaped, pendant, round apex, extends back into whitish spur as long as lip. Sepals 3, upper sepal arched as hood above column; lower sepals spreading. Stamen anthers 2, fused with style to form column above opening of spur. Ovary inferior. Fruit stalked capsule. Bract below flower. Leaf 5" long. Habitat: bogs. FL: June–September.

LOESEL'S TWAYBLADE
Liparis loeselii
Orchid family   Orchidaceae

Perennial, 6" tall, bearing several irregular green flowers on leafless stalk with 2 basal sheathing leaves. Flower $^3/_8$" long, petals 3, lower lip petal tongue-like with tip curved down; lateral petals linear and twisted. Sepals 3, upper sepal narrow, curved back; lateral sepals narrow, margins inrolled. Stamen anthers 2, fused with style to form column above stigma; beaked anther lid covers anthers. Ovary inferior, 1 style. Fruit capsule. Leaves 6" long, ovate-elongate, smooth, entire. Habitat: bogs, wet woods. FL: June–July.

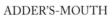

ADDER'S-MOUTH
Malaxis unifolia
Orchid family   Orchidaceae

Perennial, 8" tall, bearing spike of small irregular green flowers with flat lip and one sheathing leaf on stem. Flower $^3/_{16}$" wide, most flowers twisted upside down. Petals 3, lip petal 3-lobed, flat or edges curled under, middle lobe short; lateral petals appear tubular, slender, spreading. Sepals 3, broad dorsal sepal erect, lateral sepals spreading. Stamen anther fused to style to form column projecting over lip. Ovary inferior. Fruit capsule. Flower on twisted stalk, faces upward; subtended by bract. Leaf 3" long, oval, smooth. Habitat: damp woods, bogs. FL: July–August.

Blue to Violet

FLOWERS

AMERICAN DOG-VIOLET
Viola conspersa
VIOLET FAMILY   VIOLACEAE

Perennial, 8" tall, bearing
irregular violet flowers in axils of
leafy smooth stem. Flower stalk as
tall as or above leaves.
Flower $^3/_4$" wide, petals 5,
lower petal dark-veined and
forms violet spur, upper 2
petals reflexed wrapping
around spur, lateral violet petals
white-bearded at throat. Sepals 5,
small, pointed. Stamens 5, some
produce nectar for spur. Ovary superior,
style with some hairs. Fruit capsule. Leaf
blade $1^1/_4$" round, heart-shaped base, petioled,
blunt tip, toothed; stipule margin toothed and fringed. Habitat: damp
woods, meadows. FL: April–July.

BLUE MARSH-VIOLET
Viola cucullata
VIOLET FAMILY   VIOLACEAE

Perennial, 8" tall, bearing nodding irregular
violet flowers each on leafless stalk taller
than separate leaves. Flower $^3/_4$" wide, petals
5, upper 2 petals spreading, and lateral 2
petals bearded with white hairs.
Lower petal bearded, white base
at throat with purple veins
and spurred. Sepals 5, small,
oblong, pointed. Stamens 5,
several produce nectar for
spur. Fruit short capsule,
black seeds. Small bracts on stalk. Leaves
4" wide, long-petioled from rhizome,
toothed, smooth, pointed, heart-shaped
base, lobes often overlapped. Habitat:
marshy areas. FL: April–June.

LONG-SPURRED VIOLET
Viola rostrata
VIOLET FAMILY   VIOLACEAE

Perennial, 8" tall, bearing nodding
irregular long-spurred lavender flower
in axil on leafy smooth stem. Flower
stalk extends above leaves. Flower
$3/4$" wide, 4 petals purple-striped
at base and lower petal purple-
veined at base. Lower petal
forms $1/2$" spur with curved tip.
Calyx 5 pointed teeth. Stamens 5, some produce nectar for
spur. Ovary superior, 1 pistil, style
tip straight. Fruit $1/4$" long elliptical
capsule. Leaf blade $1^1/2$" wide, heart-
shaped, long-petioled, toothed. Stipules
large, lanceolate, toothed. Habitat: shady humus woods.
Special Concern. FL: March– July.

DOORYARD-VIOLET
Viola sororia
VIOLET FAMILY   VIOLACEAE

Perennial, 6" tall, bearing
nodding irregular purple
flowers on hairy stalks
equal or taller than
leaves. Stalks and
leaves arise from
rhizome. Flower $1/2$"
wide, petals 5, upper
2 petals spreading. Lateral
2 petals with white-bearded
base. Lower petal small, bearded
and spurred. Sepals 5, woolly, rounded
tips. Stamens 5, short. Ovary superior, short style. Fruit purple
mottled capsule. Leaf blade $1^1/2$" wide, heart-shaped, toothed, finely
hairy underside, hairy petiole. Self-pollinating flowers formed along
runners. Habitat: moist meadows, lawns. FL: April–June.

ROCK-CRESS
Arabis divaricarpa
MUSTARD FAMILY   BRASSICACEAE

Biennial, 3' tall, bearing terminal cluster of small pale violet flowers on leafy stem with hairy basal leaves. Flower $3/8$" long, petals 4, whitish bases. Sepals 4, reddish-tipped, $1/2$ length of petals. Stamens 6, two short, yellow anthers. Ovary superior, 1 pistil, short style. Fruit 3" long pod, spreading horizontally, long stalk, elongates with petals still at base. Lower leaves 2" long, oblong-lanceolate, toothed, star-shaped hairs on both surfaces. Upper leaves lanceolate, appressed to stem, basal lobes clasp stem; alternate. Habitat: sandy rocky soil. FL: April–May.

SEA-ROCKET
Cakile edentula
MUSTARD FAMILY   BRASSICACEAE

Succulent perennial, 20" tall, bearing several pale lavender flowers with yellow centers in axils of branched leafy smooth stem. Flower $1/4$" wide, petals 4, spatulate, clawed. Sepals 4, oblong, $1/2$ length of petals. Stamens 6, two short and 4 long, each with basal gland. Ovary superior, 1 pistil. Fruit short transversely 2-jointed pod with short beak, upper joint longer than lower, each with 1 seed; upper joint sheds but lower one persists on plant. Leaves 4" long, fleshy, coarsely toothed, spoon-like, taper to base; alternate. Habitat: sandy beaches. *Special Concern.* FL: July–September.

HOG-PEANUT
Amphicarpaea bracteata
PEA FAMILY   FABACEAE

Annual vine, 4' long, bearing nodding
cluster of irregular violet flowers on
short stalk in axils on hairy stem
with trifoliate leaves.
Flower $1/2$" long, corolla 2-
lipped, upper lip broad with
curled edges, lower lip short
tubular keel, and 2 small
lateral lobes narrowed to
claw. Calyx short, irregular, 4-
lobed. Stamens 10. Ovary
superior, 1 pistil, long style. Fruit
flat pod. Stem hairy, thread-like,
encircles host for support. Leaves
4" long, trifoliate, alternate. Leaflets 1" long, oval, pointed, entire,
stalked. Stipule tiny. Habitat: damp woods. FL: July–September.

CROWN-VETCH
Coronilla varia
PEA FAMILY   FABACEAE

Perennial, 2' tall, bearing umbel of
irregular violet flowers on
axillary stalk often shorter than
leaves on smooth stem with
pinnate leaves. Flower $5/8$" long,
corolla 2-lipped, upper lip broad
and oval, lower lip of 2 fused
lanceolate lobes forming keel with
curved tip, and lateral lobes
incurved. Green calyx tube broad,
3-toothed. Stamens 10.
Ovary superior, 1 pistil.
Fruit 4-angled pod. Leaves
alternate. Leaflets $3/4$" long, 8–20 pairs,
smooth, entire, short-stalked or entire. Stipule small. Habitat:
roadsides, used for erosion control; introduced. FL: June–September.

BEACH-PEA
Lathyrus maritimus
Pea family   Fabaceae

Perennial, creeping, 3'
long, bearing cluster of
irregular purple flowers
on stiff stalk shorter
than leaf on angled smooth
stem with tendril at leaf tip.
Flower 1" long, corolla 2-lipped,
upper purple lip broad, lower
white lip of 2 fused lobes
forming up-curved keel, and
lateral violet lobes narrow to
claw. Calyx irregular, 5-
pointed, with basal swelling. Stamens 10. Ovary superior, 1 pistil,
hairy style flat. Fruit long green pod. Leaves pinnate, alternate.
Leaflets 1" long, 6–12 pairs, oval, fleshy, entire, alternate. Stipule large,
arrow-shaped. Habitat: sandy lake shore. FL: June–August.

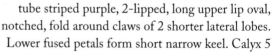

RED CLOVER
Trifolium pratense
Pea family   Fabaceae

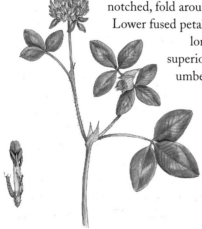

Perennial, 18" tall, bearing umbel of irregular purple flowers on short
stalk on hairy stem with trifoliate leaves. Flowers $^3/_4$" long, long corolla
tube striped purple, 2-lipped, long upper lip oval,
notched, fold around claws of 2 shorter lateral lobes.
Lower fused petals form short narrow keel. Calyx 5
long hairy teeth. Stamens 10. Ovary
superior, 1 pistil. Fruit pod. Bracts below
umbel. Leaves petioled, alternate. Oval
leaflets with yellowish chevron,
entire. Stipules striped, pointed.
Habitat: fields, pastures; introduced. FL: May–August.
White C., *T. repens,* similar but
white flower on separate stalk,
leaflets heart-shaped. Habitat:
pastures, roadsides; introduced.

AMERICAN VETCH
Vicia americana
PEA FAMILY   FABACEAE

Perennial, 3' tall,
bearing 2–9 irregular
purple flowers on stalk
shorter than leaf on smooth stem
with tendril at leaf tip. Flower
$5/8$" long, corolla 2-lipped,
upper lip broad and narrows to
claw, lower lip of 2 partly fused
short lobes forming keel,
lateral petal lobes curved
upward. Calyx lobes
unequal, finely hairy,
toothed. Stamens 10. Ovary superior, 1 pistil,
round style, tuft of hairs at tip. Fruit smooth pod. Leaves pinnate,
alternate. Leaflets 1" long, 4–7 pairs, oval, entire, stalked, alternate.
Stipule small, toothed. Habitat: moist woods. FL: May–July.

FOUR-SEEDED VETCH
Vicia tetrasperma
PEA FAMILY   FABACEAE

Annual, 2' tall, bearing 1–2 irregular violet flowers
on long slender stalk in axils on smooth stem with
tendril at leaf tip. Flower $1/4$" long, corolla 2-
lipped, upper violet lip broad
curved up and narrows to claw,
lower lip of 2 fused short lobes
forming keel, and white lateral
lobes incurved. Calyx lobe teeth
unequal, long, smooth. Stamens
10. Ovary superior, 1 pistil,
globose style tipped with
hairs. Fruit 4-seeded smooth
pod. Leaves pinnate, alternate.
Leaflets $1/2$" long, 3–5 pairs, narrow,
smooth. Stipules tiny. Habitat: cool sandy
fields; introduced. FL: May–September.

HAIRY VETCH
Vicia villosa
PEA FAMILY   FABACEAE

Perennial, 3' tall,
bearing 10–30 irregular
purple flowers on stalk longer
than leaf on hairy stem with
tendril at leaf tip. Flower $^3/_4$"
long, corolla 2-lipped, upper
blue-violet lip broad and forms
bulbous purple spur, lower violet
lip of 2 fused petals and
forms keel, and lateral
purple petals incurved. Calyx
of unequal long narrow teeth. Flowers on
1 side of stalk. Stamens 10. Ovary superior,
hair-tipped style. Fruit hairy pod. Leaves pinnate, alternate. Leaflets
$1^1/_2$" long, 5–10 pairs, linear, sessile, entire. Stipules large, deeply
toothed. Habitat: fields, roadsides; introduced. FL: June–August.

COMMON FLAX
Linum usitatissimum
FLAX FAMILY   LINACEAE

Annual, 2' tall, bearing several large
widely spreading blue flowers at
top of branches on leafy stem.
Flower $1^1/_4$" wide, petals 5,
broad, somewhat overlapping,
red-purple veins at base.
Sepals 5, narrow, small,
pointed. Stamens 5, basal
nectary, white anthers project
above petals. Ovary superior, 1 pistil,
5 styles. Fruit $^3/_8$" short-beaked capsule.
Stem branched near top, each branch forms
several flowers. Leaves $^3/_4$" long, narrow-
lanceolate, sessile; alternate. Leaves often
appressed to stem. Habitat: fields, roadsides;
introduced. FL: May–July.

BOTTLE-GENTIAN
Gentiana andrewsii
GENTIAN FAMILY   GENTIANACEAE

Perennial, 4' tall, bearing compact cluster of large oval closed blue flowers at top and in axils of unbranched leafy stem. Flower 1$^{1}/_{2}$" long, corolla with 5 short lobes separated by fringed plaits slightly longer than lobes. Calyx tubular, 5-cleft. Stamens 5, on corolla. Ovary superior, short style, lobed stigma. Fruit capsule, many seeds. Stem, leaves smooth, exude milky sap. Leaves 5" long, lanceolate, entire, pointed, sessile or slightly clasping stem; opposite. Habitat: wet meadows, shores. FL: August–October.

SOAPWORT-GENTIAN
Gentiana saponaria
GENTIAN FAMILY   GENTIANACEAE

Perennial, 4' tall, bearing compact cluster of large oval slightly open blue-violet flowers at top of leafy stem. Flower 1$^{1}/_{2}$" long, violet-striped corolla with 5 short lobes separated by fringed plaits shorter than lobes. Calyx tubular 5-cleft, short. Stamens 5. Ovary superior, 1 pistil, short style, lobed stigma. Fruit capsule, many seeds. Stem smooth, brownish-orange; exudes milky sap. Leaves 5" long, ovate, smooth, entire, pointed, pale-green underside, sessile or partially clasping; opposite. Habitat: wet woodlands. FL: August–October.

LESSER FRINGED GENTIAN
Gentianopsis procera
GENTIAN FAMILY   GENTIANACEAE

Biennial, 18" tall, bearing large fringed blue
funnel flower on long stalk at top of leafy
smooth stem. Flower 2" long, corolla
purple-striped, 4 spreading finely
toothed lobes, fringed edges, white
throat. Calyx 4 pointed lobes. Stamens
4, short, basal glands. Ovary superior,
short style. Fruit capsule. Leaves 2"
long, narrow, less than $^1/_4$" wide,
smooth, entire, partly folded, sessile;
opposite. Habitat: wet calcarious
meadows. FL: August–September.
Fringed G., *G. crinita*, similar but petal
lobes fringed, wider leaves. Habitat: wet
sandy meadows. *Special Concern.*

PERIWINKLE
Vinca minor
DOGBANE FAMILY   APOCYNACEAE

Trailing perennial, 8" tall, bearing single
tubular blue-violet flower on short stalk
in axils of leafy stem. Flower 1"
wide, corolla 5 flaring
wide lobes with white
fringed center. Calyx
5-pointed, small.
Stamens 5, within
tube. Ovaries 2,
superior, 1-celled, 2
nectaries. Fruit follicle,
few smooth seeds.
Stem forms intertwined
mats. Leaves $1^1/_2$" long,
oval-elongate, shiny, entire, sessile, dark
green; opposite. Habitat: rich woods edges,
cemeteries; introduced. FL: April–June.

BITTERSWEET NIGHTSHADE
Solanum dulcamara
<small>NIGHTSHADE FAMILY</small>   <small>SOLANACEAE</small>

Perennial vine, 10' long, bearing clusters of shooting star-shaped purple flowers with central yellow cone on short stalks in axils of leafy smooth stem. Flower $1/2$" wide, corolla 5 reflexed pointed lobes with white spots at their base. Calyx 5, small. Stamens 5, yellow anthers. Ovary superior, 1 pistil, long style. Fruit drooping red $1/4$" somewhat poisonous berry. Vine surrounds host for support. Leaves 3" long, ovate, smooth, entire, pointed, petioled, often with 2 small lobes on petiole near blade; alternate. Habitat: moist thickets, meadows; introduced. FL: July–August.

GREAT WATERLEAF
Hydrophyllum appendiculatum
<small>WATERLEAF FAMILY</small>   <small>HYDROPHYLLACEAE</small>

Perennial, 2' tall, bearing coiled cluster of bell-shaped pale lavender flowers on long hairy stalk in axils of leafy hairy stem. Flower $5/8$" long, corolla short, cleft nearly to base, 5 large spreading petal-like lobes. Calyx 5, short, narrow, hairy reflexed lobes, each alternating with reflexed appendage. Stamens 5, filaments hairy. Ovary superior, 1 forked style. Fruit capsule. Stem and stalk with short and long hairs. Upper leaves palmately lobed, petiole hairy; alternate. Lower leaves palmately 5- to 7-lobed. Habitat: moist woods, creekbeds. FL: May–June.

EASTERN WATERLEAF
Hydrophyllum virginianum
WATERLEAF FAMILY   HYDROPHYLLACEAE

Perennial, 2' tall, bearing coiled cluster of bell-shaped pale violet flowers on smooth stalk in axil of leafy stem. Flower $1/2$" long, corolla tube 5 upright lobes. Calyx 5-lobed, pointed, reflexed, finely hairy. Stamens 5, filaments purple, hairy, extend above petals. Ovary superior, 1 violet style with forked tip. Fruit capsule. Leaves 5" long, petioled, deeply pinnately lobed; alternate. Leaflets 5–7, ovate, smooth, toothed, pointed. Habitat: wet woods, creekbeds. FL: May–August.

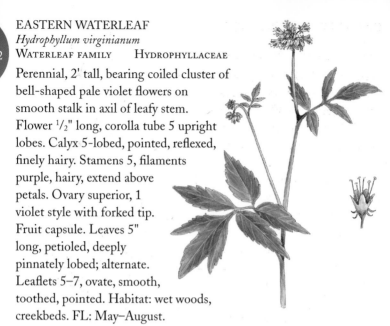

STICKSEED
Hackelia deflexa
BORAGE FAMILY   BORAGINACEAE

Biennial, 3' tall, bearing funnel-shaped pale blue cream-eyed flowers on forked branches at top of leafy hairy stem. Flower $1/8$" wide, corolla 5 flared lobes, cream eye or corona formed from corolla scales at tube throat. Calyx 5-toothed, hairy. Stamens 5. Ovary superior, short stigma-style. Fruit 4 nutlets, outer surface of each nutlet covered with 10–20 hooked hairs (below). Stem tips uncoil during growth. Lower leaves 6" long, lanceolate, downy, stiff hairs on underside; alternate. Upper leaves downy, shorter, become sessile among flowers. Habitat: rich dry woods. FL: May–August.

GARDEN FORGET-ME-NOT
Myosotis sylvatica
BORAGE FAMILY   BORAGINACEAE

Perennial, 18" tall, bearing funnel-shaped
blue yellow-eyed flowers on branches at top
of leafy hairy stem. Flower $1/4$" wide,
corolla 5-lobed, yellow eye or corona
formed from corolla scales at tube
throat. Calyx 5-toothed,
hooked hairs. Stamens 5.
Ovary superior, short style.
Fruit hairy nutlets; style longer
than nutlet. Stem tip uncoils
during growth. Leaves 2" long,
ovate, hairy, entire, sessile; alternate.
Habitat: moist rich woods; introduced.
FL: May–October.
Smaller F-m-n., *M. laxa*, leaves blunt-
tipped, calyx hairs not hooked, style shorter
than nutlet. Habitat: streamside.

COMMON VERVAIN
Verbena hastata
VERVAIN FAMILY   VERBENIACEAE

Perennial, 4' tall, bearing narrow branched
candelabra spikes of crowded slightly irregular
violet flowers at top of leafy hairy square
stem. Flower $1/8$" wide, long curved corolla
tube, 5 widely spreading lobes. Calyx 5-lobed,
small, purple, pointed. Stamens 4, in
corolla tube. Ovary superior, short
bilobed stigma-style. Fruit dry, 4
nutlets. Short bract below flower.
Stem ridged. Leaves 6" long, narrow-
ovate, toothed, pointed,
rough, petioled or clasping
stem; opposite. Lower leaves
3-lobed. Habitat: wet meadows,
lakeshores. FL: June–October.

HEMP-NETTLE
Galeopsis tetrahit
<small>MINT FAMILY</small>   <small>LAMIACEAE</small>

Annual, 18" tall, bearing
dense cluster of irregular
violet flowers in upper
axils of branched square
leafy stem with reflexed
hairs. Flower ³/₈" long,
corolla tube hairy 2-
lipped, upper lip
hooded, lower lip 3-
lobed with 2 basal
protuberances. Calyx
tubular, 5-toothed, finely
hairy. Stamens 4, woolly, under
hood. Ovary superior, 1 pistil, 4-lobed,
styles equal. Fruit smooth nutlets. Leaves 4"
long, ovate, toothed, finely hairy, petioled; opposite. Habitat: rocky
shores, rich woods; introduced. FL: July–September.

SPEARMINT
Mentha spicata
<small>MINT FAMILY</small>   <small>LAMIACEAE</small>

Perennial, 3' tall, bearing elongated cluster
of fragrant irregular pale violet flowers at
top of leafy smooth square stem. Flower
³/₁₆" long, corolla tube 2-lipped, each lip
2-lobed, all nearly equal. Calyx weakly 2-
lipped, 4-pointed, hairy. Stamens 4,
protruding above corolla. Ovary
superior, 1 pistil, style protruding above corolla. Fruit 4
nutlets. Leaves 2¹/₂" long, ovate,
toothed, smooth but hairy
midrib on underside, sessile,
spearmint-scented; opposite.
Habitat: roadsides, fields; introduced. FL: July–September.

WILD BERGAMOT
Monarda fistulosa
MINT FAMILY   LAMIACEAE

Perennial, 3' tall, bearing large cushion
of fragrant irregular tubular violet
flowers at top of leafy hairy square
stem. Flower 1" long, corolla tube 2-
lipped, upper lip 2-lobed, hairy; lower
lip 3-lobed. Calyx tube regular, short,
woolly. Stamens 2, projecting from
tube. Ovary superior, 1 pistil, style
projecting from tube. Fruit nutlet.
Cushion of flowers subtended by leafy
and hairy bracts; calyxes persistent.
Stem may be branched. Leaves 2$^1/_2$"
long, ovate, toothed, petioled;
opposite. Habitat: dry wood thickets,
meadows. FL: June–September.

SELF-HEAL
Prunella vulgaris
MINT FAMILY   LAMIACEAE

Perennial, 1' tall, bearing spikes of irregular violet
flowers at top and in axils of leafy square
stem. Flower $^1/_2$" long, corolla tube 2-
lipped, upper lip arched as hood, lower
lip 3-lobed, broad white lobe fringed
along edge. Calyx tube 2-lipped,
hairy, persistent on spike. Stamens
4, under hood. Ovary superior,
deeply 4-lobed, 1 pistil, forked long
style. Fruit 4 nutlets. Bract
beneath each flower
sharply tipped. Leaves 3"
long, elliptic, pointed,
smooth, several teeth, entire, petioled to
entire; opposite. Habitat: roadsides, fields,
disturbed areas; introduced. FL: June–August.

MARSH-SKULLCAP
Scutellaria galericulata

MINT FAMILY   LAMIACEAE

Perennial, 2' tall, bearing single large irregular
upturned light blue-violet flower in axils
near top of leafy square stem. Flower
$3/4$" long, corolla tubular 2-lipped,
upper lip smooth or notched,
hood-like, lower lip 3-lobed
middle dotted. Calyx tube short,
2-lipped with conical protuberance on upper lip. Stamens 4, in
throat. Ovary superior, 1 pistil,
deeply 4-lobed, 1 style. Fruit 4 white
nutlets. Stem finely hairy at angles,
recurved hairs. Leaves 2" long, ovate,
toothed, smooth, pointed, short-petioled;
opposite. Habitat: wet shady woods, along
streams. FL: July–August.

MAD-DOG SKULLCAP
Scutellaria lateriflora

Mint family   Lamiaceae

Perennial, 2' tall, bearing several irregular
violet flowers in succession only on
one side of stalk from axils near top of
leafy square stem. Flower $3/16$" wide,
corolla tube 2-lipped, nearly straight.
Calyx short, 2-lipped, upper lip
bearing protuberance. Stamens 4,
in upper lip. Ovary superior,
deeply 4-lobed, 1 pistil, 1 style.
Fruit 4 green nutlets (below).
Stem angles may be minutely
hairy. Leaves longest at middle
of stem; petiole length increases
toward stem base. Leaves $1^1/2$" long, ovate,
coarsely toothed, smooth; opposite. Habitat:
wet woods, marshes. FL: July–August.

MONKEY-FLOWER
Mimulus ringens
FIGWORT FAMILY   SCROPHULARIACEAE

Perennial, 3' tall, bearing single irregular violet flower in axils on leafy square stem with musk scent. Flower $3/4$" wide, corolla tube 2-lipped, upper lip 2 curled lobes, lower lip of 3 scalloped lobes with pink spots; white flower throat is closed by palate. Stamens 4. Ovary superior, long style, 2-lobed stigma. Fruit cylindrical capsule, black-tipped seeds. Flower stalk about twice length of calyx. Stem glandular, branched. Leaves 4" long, lanceolate, pointed, toothed, sessile or clasping stem; opposite. Habitat: stream banks, wet meadows. FL: June–September.

AMERICAN SPEEDWELL
Veronica americana
FIGWORT FAMILY   SCROPHULARIACEAE

Perennial, 3' tall, bearing spikes of blue-violet flowers in axils of leafy smooth stem. Flower $3/8$" wide, corolla 4-lobed, purple striped, white ring and green center at throat; lower lobe small. Calyx 4-lobed, pointed, short. Stamens 2, white anthers project above petals. Ovary superior, long style. Fruit $1/8$" wide, flat, turgid capsule with beak; calyx lobes shorter than height of capsule. Flower stalk longer than subtending bract. Lower leaves oval, entire, smooth, petiole clasping; opposite. Upper leaves narrow-ovate, coarsely toothed, petiole clasping. Habitat: shallow water. FL: May–September.

WATER-SPEEDWELL
Veronica anagallis-aquatica
<small>FIGWORT FAMILY   SCROPHULARIACEAE</small>

Perennial, 3' tall, bearing spikes
of irregular violet flowers in
axils of leafy smooth stem.
Flower $^3/_8$" wide, corolla 4
spreading lobes, purple-striped,
lower lobe small; white ring
with green center at throat.
Calyx 4-lobed, short, hairy,
persistent. Stamens 2, dark
anthers above petals. Ovary superior, 1
pistil, long style. Fruit heart-shaped, flat,
turgid capsule, persistent style; calyx as long as
capsule. Flower stalk $^3/_8$" long, longer than
bract. Leaves 2" long, lanceolate, pointed,
toothed, smooth, sessile or clasping; opposite.
Habitat: swamps, wet ditches; introduced. FL: May–September.

GERMANDER-SPEEDWELL
Veronica chamaedrys
<small>FIGWORT FAMILY   SCROPHULARIACEAE</small>

Perennial, 8" tall, bearing clusters of small
irregular blue flowers with central white
eye on long stalks in axils on finely
hairy leafy stems. Flower $^3/_8$" wide,
corolla 4 spreading lobes, lower lobe
narrower than others, all
with blue veins; white
ring at throat. Calyx 4,
small teeth, hairy. Sta-
mens 2, whitish anthers project
above petals. Ovary superior, long
style. Fruit $^1/_8$" wide, flat, heart-shaped,
hairy capsule, persistent style. Bract of
flower stalk shorter than stalk. Leaves 1"
long, ovate, coarsely toothed, mostly sessile;
opposite. Habitat: lawn, fields; introduced. FL: May–July.

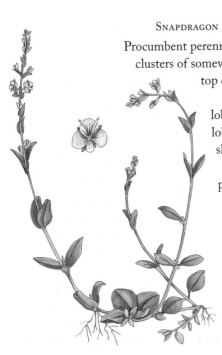

COMMON SPEEDWELL
Veronica officinalis
<small>FIGWORT FAMILY   SCROPHULARIACEAE</small>

Perennial, 10" tall, bearing loose spikes of violet flowers in axils of leafy finely hairy stem. Flower ³/₈" wide, corolla tube 4 spreading lobes, purple striped, lower lobe small. Calyx deeply 4-lobed, hairy, ¹/₃ length of petals. Stamens 2, dark anthers above corolla. Ovary superior, long style. Fruit ¹/₈" wide, heart-shaped, flat capsule, glandular, persistent style. Bract subtending flower stalk as long as stalk. Leaves 2" long, oval, very short petiole, toothed, velvety; opposite. Habitat: upland woods, fields; introduced. FL: May–July.

THYME-LEAVED SPEEDWELL
Veronica serpyllifolia
<small>SNAPDRAGON FAMILY   SCROPHULARIACEAE</small>

Procumbent perennial, 8" tall, bearing pale small clusters of somewhat irregular violet flowers at top of short upturned leafy stems. Flower ¹/₄" long, corolla 4-lobed, purple stripes on 3, small lobe not striped. Calyx 4-lobed, shorter than petals. Stamens 2, white anthers extend above petals. Ovary superior, 1 pistil, long style. Fruit minutely hairy, flat, heart-shaped capsule, persistent style. Prostrate stem branched, forms mats. Leaves ¹/₂" long, oval, entire, sessile, smooth: opposite. Habitat: lawns, disturbed soil; introduced. FL: April–August.

CANCER-ROOT
Orobanche uniflora
BROOM-RAPE FAMILY   OROBANCHACEAE

Perennial, 6" tall, bearing single fragrant irregular pale-violet flower at top of finely hairy cream stalk. Flower $^3/_4$" long, corolla tube curved with 5 spreading violet striped lobes, lower lobe pointed and longer than other 4 round-tipped lobes. Calyx cup-shaped, $^1/_3$ length corolla, 5 pointed lobes. Stamens 4, dark yellow anthers at white throat of corolla tube. Ovary superior, long style, large stigma. Fruit capsule, many small seeds. Underground stem produces 1–several stalks with single flower. Leaves small cream scales at stalk base. Habitat: moist sandy woods, shores. *Special Concern.* FL: May-June.

MARSH-BELLFLOWER
Campanula aparinoides
BELLFLOWER FAMILY   CAMPANULACEAE

Perennial, 2' tall, bearing solitary bell-shaped violet flower at top of floppy branched hairy stem with narrow leaves and milky sap. Flower $^1/_4$" wide, corolla cleft over half length into 5 spreading lobes. Calyx 5-lobed, short. Stamens 5, white anthers. Ovary inferior, short style. Fruit globose capsule, many seeds. Stem slender, 3-angled, reflexed bristles along angles (below). Leaves 1" long, very narrow, appear sessile, rough margin and midrib; alternate. Habitat: marshes, swales. FL: June–August.

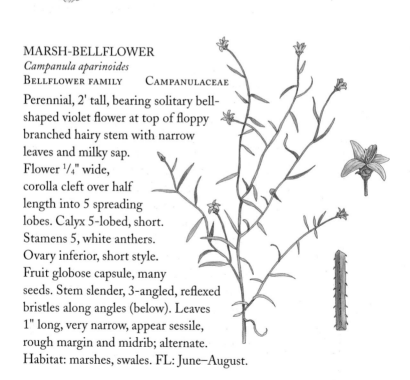

HAREBELL
Campanula rotundifolia
BELLFLOWER FAMILY   CAMPANULACEAE

Perennial, 2' tall, bearing nodding bell-shaped blue-violet flower on very thin stalk in axils near top of smooth leafy stem. Flower $^3/_4$" long, corolla 5-lipped. Calyx 5-pointed, short, spreading. Stamens 5. Ovary inferior, 1 pistil, 3-lobed style. Fruit capsule, many seeds. Stem branched near top, smooth; exudes milky sap. Basal leaves round, petioled; may be absent. Stem leaves 2" long, narrow, smooth, sessile, entire; alternate. Habitat: sandy, gravelly well-drained soil, lake shores. FL: June–August.

BROOK LOBELIA
Lobelia kalmii
BELLFLOWER FAMILY   CAMPANULACEAE

Perennial, 18" tall, bearing several irregular light-blue white-throated flowers on slender stalk in axils of leafy stem. Flower $^1/_2$" long, corolla tube 2-lipped, lower lip 3-lobed with white markings and the erect smooth upper lip 2-lobed. Calyx 5 very thin teeth, short. Stamens 5, encircle style. Ovary inferior, 1 pistil, style lobed. Flower upside-down on twisted stalk. Stem, leaves exude acrid milky sap. Basal leaves, if still present, spatulate. Stem leaves 2" long, linear-narrow, smooth or finely toothed; alternate. Habitat: calcareous wet areas, wet roadside, creeksides. FL: July–August.

GREAT LOBELIA
Lobelia siphilitica

BELLFLOWER FAMILY   CAMPANULACEAE

Perennial, 4' tall, bearing cluster of large irregular purple flowers at top of leafy hairy stem with milky sap. Flower 1" long, corolla tube 2-lipped, upper lip with 2 erect narrow lobes and lower lip with 3 wide down-curved lobes. Calyx 5-pointed. Stamens 5, encircle green style, both protrude through slit in corolla and terminate between lobes of upper lip. Ovary inferior. Fruit capsule. Flower on short stalk, subtended by 1–2 bracts. Leaves 3" long, lanceolate, toothed, pointed, short petiole to sessile, finely hairy; alternate. Habitat: swamps, creeksides. FL: August–September.

TEASEL
Dipsacus laciniatus

TEASEL FAMILY   DIPSACACEAE

Biennial, 4' tall, bearing irregular tubular dusty violet flowers in thistle-like round head subtended by armed bracts on spiny leafy ridged stem. Flower ¼" long, corolla tube 4-lobed. Calyx 4-lobed, short, hairy. Stamens 4, extend above corolla. Ovary inferior, style extends above corolla. Fruit achene. Bracts among flowers awl-pointed, length of flower. Long prickly bracts at base of head, curve up like cup as flowering progresses. Basal leaves 10" long, lanceolate, spiny. Stem leaves 8" long, pinnately lobed, perfoliate, prickly; opposite. Habitat: disturbed areas; introduced. FL: July–August

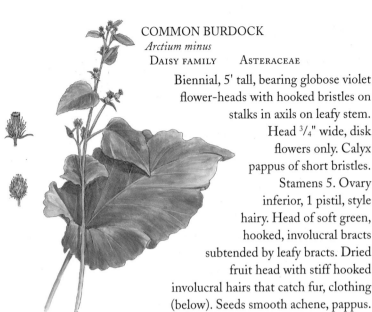

COMMON BURDOCK
Arctium minus
<small>DAISY FAMILY   ASTERACEAE</small>

Biennial, 5' tall, bearing globose violet flower-heads with hooked bristles on stalks in axils on leafy stem. Head ³/₄" wide, disk flowers only. Calyx pappus of short bristles. Stamens 5. Ovary inferior, 1 pistil, style hairy. Head of soft green, hooked, involucral bracts subtended by leafy bracts. Dried fruit head with stiff hooked involucral hairs that catch fur, clothing (below). Seeds smooth achene, pappus. Lower leaves 18" long, ovate, heart-shaped base, whitish underside scalloped, petioled; alternate. Upper leaves smaller. Habitat: disturbed areas; introduced. FL: July–October.

COTTON-BURDOCK
Arctium tomentosum
<small>DAISY FAMILY   ASTERACEAE</small>

Biennial, 4' tall, bearing cottony globose purple flower-heads on long stalks from upper axils on leafy stem. Head 1" wide, central cluster of purple disk flowers only. Corolla glandular outside. Calyx pappus of short bristles. Stamens 5, filaments separate. Ovary inferior, 1 pistil, style hairy at base of branch. Fruit smooth flat oblong achene, pappus. Involucral bracts many, fringed with bluish-gray cottony hairs, head appears puffy. Leaves 18" long, ovate, smooth above, fuzzy underside, entire, petioled; alternate. Habitat: moist disturbed areas; introduced. FL: June–October.

NORTHERN HEART-LEAVED ASTER
Aster ciliolatus

DAISY FAMILY   ASTERACEAE

Perennial, 3' tall, bearing finely hairy
inflorescence of many pale violet flower-
heads on stem with winged heart-shaped
leaves. Head ⅝" wide, 10–20 ray, and
several yellow disk flowers. Calyx pappus of
hairs. Ovary inferior, forked style. Fruit
smooth achene, pappus. Few leafy bracts on
flower stalks. Involucral bracts narrow,
pointed, smooth or ciliate margin,
green tip. Lower leaves 4" long,
toothed, pointed, finely hairy
underside, winged petiole, leaves fall off
early. Upper leaves ovate, winged petiole,
entire; alternate. Habitat: wood edges, clearings. FL: August–October.

BIG-LEAVED ASTER
Aster macrophyllus

DAISY FAMILY   ASTERACEAE

Perennial, 2' tall, bearing branched
terminal cluster of large violet flower-
heads on leafy glandular stem with
heart-shaped leaves. Head 1¼"
wide, 10–16 ray, and yellow
disk flowers. Calyx pappus of
capillary hairs. Ovary
inferior, 1 pistil, forked style
hairy. Fruit ribbed achene,
pappus. Leafy bract below
each head stalk. Involucral
bracts pointed, overlapping,
glandular. Basal and lower stem
leaves 8" long, long-petioled, rough, toothed.
Upper leaves smaller, pointed, small-toothed,
winged petiole; alternate. Habitat: rich woods;
colonial. FL: July–October.

NEW ENGLAND ASTER
Aster novae-angliae
DAISY FAMILY   ASTERACEAE

Perennial, 4' tall, bearing branched
terminal cluster of large purple-violet
flower-heads at top of leafy glandular
stem with clasping leaves. Head 2" wide,
40–50 narrow ray, and many yellow disk
flowers. Calyx pappus of hairs. Ovary
inferior, 1 pistil, forked style. Fruit ribbed
finely hairy achene, pappus. Involucral bracts
green, reflexed, pointed, hairy, glandular,
overlapping, red-tipped. Upper stem red.
Leaves 5" long, hairy, lanceolate, entire,
lower sessile, lobed base of upper leaves
clasp hairy stem; alternate. Habitat: moist meadows.
FL: August–September.

AWL-ASTER
Aster pilosus
DAISY FAMILY   ASTERACEAE

Perennial, 4' tall, bearing
inflorescence of pale violet
flower-heads on leafy hairy stem
with linear leaves. Head $^3/_4$" wide,
15–30 ray, and yellow disk flowers.
Calyx pappus of hairs.
Ovary inferior, forked
style. Fruit smooth achene,
pappus. Involucre urn-shaped; bracts
overlapping, green, pointed.
Axillary floral stalks with short
sessile leaves; stalks subtended by persistent
leaf longer than
flower stalk.
Leaves linear $3^1/_2$" long × $^3/_8$" wide, sessile,
entire, pointed. Basal and lower stem leaves often twisted, dry
and fall early. Habitat: dry sandy soil. FL: August–October.

BRISTLY ASTER

Aster puniceus

<small>DAISY FAMILY</small>   <small>ASTERACEAE</small>

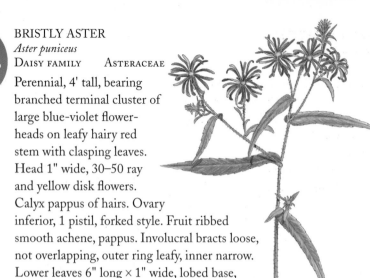

Perennial, 4' tall, bearing
branched terminal cluster of
large blue-violet flower-
heads on leafy hairy red
stem with clasping leaves.
Head 1" wide, 30–50 ray
and yellow disk flowers.
Calyx pappus of hairs. Ovary
inferior, 1 pistil, forked style. Fruit ribbed
smooth achene, pappus. Involucral bracts loose,
not overlapping, outer ring leafy, inner narrow.
Lower leaves 6" long × 1" wide, lobed base,
lanceolate, lower toothed. Upper leaves narrow,
pointed, toothed, tend to clasp stalk; alternate. Habitat: swamps,
wet woods. FL: August–October.

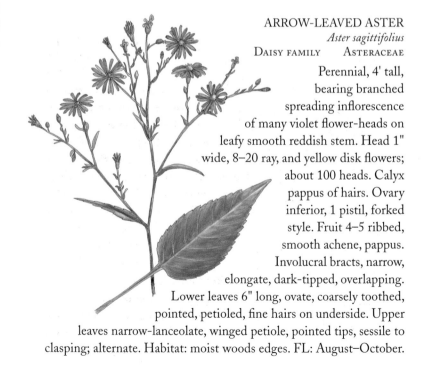

ARROW-LEAVED ASTER

Aster sagittifolius

<small>DAISY FAMILY</small>   <small>ASTERACEAE</small>

Perennial, 4' tall,
bearing branched
spreading inflorescence
of many violet flower-heads on
leafy smooth reddish stem. Head 1"
wide, 8–20 ray, and yellow disk flowers;
about 100 heads. Calyx
pappus of hairs. Ovary
inferior, 1 pistil, forked
style. Fruit 4–5 ribbed,
smooth achene, pappus.
Involucral bracts, narrow,
elongate, dark-tipped, overlapping.
Lower leaves 6" long, ovate, coarsely toothed,
pointed, petioled, fine hairs on underside. Upper
leaves narrow-lanceolate, winged petiole, pointed tips, sessile to
clasping; alternate. Habitat: moist woods edges. FL: August–October.

BROWN KNAPWEED
Centaurea jacea
DAISY FAMILY   ASTERACEAE

Perennial, 3' tall, bearing single violet flower-
head at tips of several branches on smooth leafy
stem. Head 1¹/₄" wide, disk flowers only.
Disk corolla of flowers along
receptacle edge modified into
elongated forked red-violet
"petals." Central disk flowers less
modified. Calyx pappus absent.
Stamens 5. Ovary inferior, 1 pistil,
style hairy. Fruit smooth achene.
Involucral bracts brown, jagged
tips. Lower leaves 4" long, elongated, smooth, toothed. Upper leaves
narrow-lanceolate, toothed or entire,
sessile; alternate. Habitat: moist roadsides,
fields; introduced. FL: June–September.

SPOTTED KNAPWEED
Centaurea maculosa
DAISY FAMILY   ASTERACEAE

Bi-perennial, 3' tall, bearing
single prickly violet flower-
head at top and in axils of
branches of finely hairy leafy stem.
Head 1" wide, disk flowers only.
Corolla modified into narrow elongated
and sometimes forked violet "petals."
Calyx pappus of short bristles.
Stamens 5. Ovary inferior, 1 pistil,
style hairy at base. Fruit smooth
achene, pappus. Involucral bracts
prickly, toothed, brown-tipped. Leaves
8" long, pinnate, leaflets narrow, much
dissected into narrow segments, gray-green;
upper leaves smaller. Habitat: roadsides, fields;
introduced. FL: June–August.

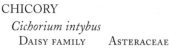

CHICORY

Cichorium intybus

DAISY FAMILY   ASTERACEAE

Perennial, 4' tall, bearing 1–3 blue flower-heads in axils along leafy hairy stem with milky sap. Head 1" wide, 16 or more ray flowers only. Corolla square-tipped, 5-notched. Calyx pappus of minute scales. Stamens 5, encircle style. Ovary inferior, 1 pistil, forked style. Fruit 5-angled achene, pappus. Involucral bracts 8, smooth. Head can be nearly sessile to stem. Flower heads open 1 day only. Basal leaves pinnately lobed, toothed, dandelion-like, 6" long, hairy underside. Upper leaves smaller, narrow, tend to clasp stem; alternate. Habitat: roadsides, fields; introduced. FL: June–September.

CANADA-THISTLE

Cirsium arvense

DAISY FAMILY   ASTERACEAE

Perennial, 5' tall, bearing many large fragrant pale lavender flower-heads in flat-topped inflorescence on leafy smooth stem. Head 1" wide, disk flowers only. Corolla tubular. Calyx pappus of bristles. Stamens 5. Ovary inferior, 1 pistil, style lobed, hairy. Fruit flat, veined achene, pappus. Involucre $^3/_4$" high, weakly spiny bracts. Flower-head subtended by leafy bract. Leaves 8" long, spiny margin, somewhat pinnately lobed, finely hairy underside, petiole clasps stem: alternate. Habitat: pastures, disturbed areas, in colonies; introduced. FL: June–October.

BULL-THISTLE
Cirsium vulgare
DAISY FAMILY   ASTERACEAE

Biennial, 5' tall, bearing many large fragrant spiny red-violet flower-heads on leafy spiny stem. Head 1¼" wide, disk flowers only. Long violet corolla white-tipped. Calyx pappus of bristles. Stamens 5. Ovary inferior, 1 pistil, lobed hairy style. Fruit flat ribbed achene, pappus. Involucre 1" high, urn-shaped, spiny bracts. Leaves 6" long, deeply lobed with spiny tips, margins with sharp teeth. Leaf base with wings that extend onto and down stem, wings lobed and spiny. Habitat: fields, pastures; introduced. FL: June–September.

DWARF LAKE-IRIS
Iris lacustris
IRIS FAMILY   IRIDACEAE

Perennial, 6" tall, bearing large purple flower on stalk with flat sheathing leaves. Flower 1½" wide, 3 purple sepals with yellow crest surrounded by white fringe and dark-purple markings. Three purple petals with notched tip. Three elongated purple styles with forked curled tip are above sepals; stigmas hidden under curled tip. Three stamens attached to sepals, but hidden under style. Ovary inferior; fruit capsule, ariled seeds. Bracts below flower. Leaves 6" long, broad, curved, arching. Habitat: calcareous soil along Lake Michigan shore. *Threatened.* FL: May.

LARGER BLUE FLAG
Iris versicolor
IRIS FAMILY   IRIDACEAE

Perennial, 3' tall, bearing large purple flowers with
yellow-marked sepals on leafy stalk. Flower 2" wide,
3 purple sepals with bright yellow crest surrounded
by white area with purple veins; clawed. Three
light purple petals, notched tip; clawed. Three
light purple styles with forked curled tip over
sepals; stigma under curled tip. Three stamens
attached to sepals; hidden under style. Ovary
inferior, fruit 3-angled capsule. Bracts below
flower. Basal leaves, $3/8$" wide, linear, shorter than
stalk, folded. Habitat: wet meadows. FL: May–July.
Southern B. F., *I. virginica*, leaves longer than flower
stalk. Habitat: marshes, swamps.

BLUE-EYED GRASS
Sisyrinchium montanum
IRIS FAMILY   IRIDACEAE

Perennial, 20" tall, bearing single
cluster of blue-violet flowers with
yellow centers near top of a flat twisted
stem. Flower 1" wide, 6 tepal, each
pointed. Stamens 3, filaments
united to form column
around style. Ovary
inferior, 1 pistil, style 3-
lobed. Fruit capsule, black
wrinkled seeds. Flowers subtended and overtopped by long
bract. Flower stem flat, sharp-edged,
twisted. Basal leaves grass-like, folded,
length varies. Habitat: fields, moist
meadows. FL: May–July.

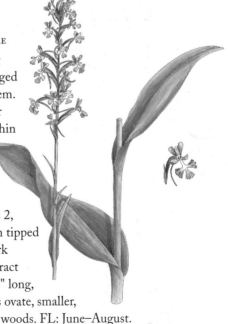

HELLEBORINE ORCHID
Epipactis helleborine
ORCHID FAMILY   ORCHIDACEAE

Perennial, 2' tall, bearing terminal spike of
irregular nodding red-violet and green
flowers on leafy stem. Flower $^1/_2$" wide,
petals 3, lower petal forms inflated sac;
upper 2 petals oval-lanceolate, spreading, arched over column. Sepals 3, oval,
upper light-green sepal arched over
column; lateral violet green sepals spreading.
Stamen anthers 2, fused with style to form
column that projects over sac; stigma under
anthers. Ovary inferior, 1 style. Fruit
capsule, many seeds. Linear
smooth bract subtends flower.
Leaves 5" long, ovate, clasp
stem, smooth; alternate. Habitat: rich
woods, gardens; introduced. FL: June–August.

PURPLE FRINGED ORCHID
Habenaria psycodes
ORCHID FAMILY   ORCHIDACEAE

Perennial, 30" tall, bearing long
cluster of fragrant irregular fringed
purple flowers at top of leafy stem.
Flower $^1/_2$" wide, petals 3, lower
petal 3 fringed lobes and long thin
purple spur, other petals
erect often notched. Sepals
3, oval, upper sepal forms hood
over column, other two sepals
spread laterally. Stamen anthers 2,
fused with style to form column tipped
with stigma. Ovary inferior, dark
purple. Fruit capsule. Narrow bract
subtends flower. Lower leaves 8" long,
ovate, sheath stem; upper leaves ovate, smaller,
clasp stem. Habitat: cool moist woods. FL: June–August.

Bibliography

Courtenay, Booth, and James H. Zimmerman, *Wildflowers and Weeds*. New York: Prentice Hall Press, 1972.

Cox, Donald D., *Common Flowering Plants of the Northeast*. Albany: State University of New York Press, 1985.

Eggers, Steve D., and Donald M. Reed, *Wetland Plants and Plant Communities of Minnesota & Wisconsin*. 2nd ed. St. Paul: U. S. Army Corps of Engineers,1997.

Fassett, Norman C., *Spring Flora of Wisconsin*. 3rd ed. Madison: University of Wisconsin Press, 1957.

Fernald, Merritt L., *Gray's Manual of Botany*. 8th ed. New York: American Book Company, 1950.

Fuller, Albert M., *Studies on the Flora of Wisconsin. Part I: The Orchids; Orchidaceae*. Bull. Public Mus. City Milw. 14(1):1–284, 1993.

Gleason, Henry A., and Arthur Cronquist, *Manual of Vascular Plants of Northeastern United States and Adjacent Canada*. 2d ed. Bronx: New York Botanical Garden, 1991.

Henn, Robert H., *Wildflowers of Ohio*. Bloomington: Indiana University Press, 1998.

Luer, Carlyle A., *The Native Orchids of the United States and Canada*. Bronx: New York Botanical Garden, 1975.

Newcomb, Lawrence, *Newcomb's Wildflower Guide*. Boston: Little, Brown and Co., 1977.

Niering, William A., and Nancy C. Olmstead, *National Audubon Society Field Guide to North American Wildflowers*. New York: Alfred A. Knopf, 1995.

Peterson, Roger T., and Margaret McKenny, *A Field Guide to Wildflowers*. Boston: Houghton Mifflin, 1968.

Rabeler, Richard K., *Gleason's Plants of Michigan*. 1st ed. Ann Arbor: Oakleaf Press, 1998.

Index

Artist MARILYN WAITE MAHLBERG
and botanist PAUL G. MAHLBERG live in Bloomington,
Indiana, during the academic year, and in Door County,
Wisconsin, in the summer. Their love of Door County
inspired the Mahlbergs to locate and write about (Paul) and
portray in ink and watercolor (Marilyn) the approximately
380 wildflowers included in this field guide.

Book and Cover Designer: Sharon L. Sklar

Copy Editor: Bobbi Diehl

Compositor: Sharon L. Sklar

Typefaces: Adobe Caslon and Twang

Book and Cover Printer: Four Colour Imports